INTRODUCTION

This medal roll has been transcribed from a hand written card index system, formerly located at the Air Ministry, which contained details of both the *India General Service Medal 1908-1935* and the subsequent issue, *India General Service Medal 1936-1939.*

Where a recipient was also entitled to the later medal, a note has been made in the Comments section.

* * * * * *

CLASPS

Qualification

The Royal Air Force qualified for the following clasps:-

Campaign	Clasp No. on Roll
Afghanistan NWF 1919	1
Mahsud 1919-20	2
Waziristan 1919-21	3
Waziristan 1921-24	4
Waziristan 1925	5
NW Frontier 1930-31	6
Burma 1930-32	7
Mohmand 1933	8
NW Frontier 1935	9

Clasp Nos. 1-6 were first issued with the 2nd obverse of this medal, clasp Nos. 7-9 were first issued with the 3rd obverse.

Analysis

Campaign	Squadrons Present
Afghanistan NWF 1919	20,31,48,114
Mahsud 1919-20	20,31,48,114
Waziristan 1919-21	5,20,27,28,31,60,97*,99*
Waziristan 1921-24	5,27,28,31,60
Waziristan 1925	5,27,60
NW Frontier 1930-31	5,11,20,27,28,39,60
Burma 1930-32	36,205
Mohmand 1933	20
NW Frontier 1935	20

+ These Squadrons formed 52 Wing.
* One Flight only.

Campaign	Clasps Issued Definite	Possible	Total
Afghanistan NWF 1919	803	55	858
Mahsud 1919-20	162	7	169
Waziristan 1919-21	483	109	592
Waziristan 1921-24	581	6	587
Waziristan 1925	280	61	341
NW Frontier 1930-31	1356	12	1368
Burma 1930-32	13	1	14
Mohmand 1933	180	1	181
NW Frontier 1935	930	1	931
Clasps undefined			26
TOTAL			5067

Notes

Waziristan 1925

This clasp was awarded for service under Wing Commander R.C.M. Pink C.B., against the Wazaris. It was awarded **only** to the Royal Air Force and this was the first occasion in which the RAF was used as an independent force.

A.M.W.O. 643/26 of 2nd December 1926 stated that those who had previously qualified for the Waziristan 1921-24 clasp and who took part in the 1925 operations in Waziristan, would be given the option of receiving **either** the Waziristan 1921-24 clasp or the Waziristan 1925 clasp. They would not be issued with both.

A review of both official and unofficial publications reveal the differences in the number of Waziristan 1925 clasps apparently authorised and issued. Readers should consult:-

London Gazette 20th November 1925

OMRS Journal (September 1968) - S/L NB Pickstock

British Battles & Medals - Spink & Son 1988

OMRS Journal (Winter 1990) - W/C J Routeledge for further information.

Abbreviations

The following abbreviations are used in the medal roll:-

( ) - No clasp is defined on the record card.

(5) - Entitlement to the clasp shown in brackets cannot be confirmed from the record card. In some cases, reference to other official documents can verify the clasp entitlement.

N.R.V.I. - There is no record that the clasp/medal was verified and/or issued.

Medal O.C. - The record card notes that the medal has been officially corrected.

Printed and bound in Great Britain by
Antony Rowe Ltd, Eastbourne

Name	Rank	No.	Clasp	Comments
Abbey, A.H.	AC1	326462	4	
Abbotts, S.	AC2	508621	6	
Abbs, W.H.	AC2	336654	(3),6	28 Squadron
Abernethy, J.M.	AC1	33673	(3)	28 Squadron
Abraham, R.B.	F/O		6	
Abraham, T.M.	F/O		9	Also 36/37 as S/L
Absom, E.	Cpl	9870	1	
Ace, W.A.A.	AC1	513209	9	
Adam, M.J.	F/O		6,9	
Adams, C.	Cpl	345329	5,9	
Adams, C.P.	LAC	513097	9	Also 36/37 as Cpl
Adams, H.C.	WO1	3337	(3)	3rd Wing
Adams, H.E.	LAC	332899	(3)	
Adams, J.	AC1	510109	9	
Adams, J.J.	LAC	505532	6	
Adams, J.S.L.	F/O		9	
Adams, L.C.	LAC	366202	9	
Adams, R.	A/Cpl	127233	1	
Adams, R.	AC2	338081	2,3	
Adams, R.A.	AC1	515325	9	Also 36/37
Adamson, C.W.A.	LAC	358518	6	
Adamson, T.A.H.	AC1	245537	2,3	
Adderley, Mr, A.	*		(9)	* Education Officer. N.R.I.V.
Addis, T.H.	Sgt	243688	6	
Agnew, H.	AM2	22302	1	
Ahern, W.G.	Sgt	363525	6	
Aiken, L.W.	F/O		5	Deceased
Airey, J.L.	F/L		8,9	
Aitken, A.	Cpl	239993	9	
Aitken, A.	AC1	513170	8,9	
Alcock, R.	AM2	36672	1	
Alderson, F.H.	LAC	560522	9	
Alderton, N.S.	AM1	409154	1	
Aldir, P.J.	LAC	512874	9	
Aldred, G.C.	AMZ	404259	1,2,3	
Allored, T.	AC1	359303	6	
Aldridge, S.R.	AC1	356202	6	
Alexander, H.	O/O		1	48 Squadron
Alexander, W.C.	LAC	359884	(6)	N.R.I.V.
Alexander, W.F.	AC2	508826	6	
Alford, E.G.	F/Sgt	335383	9	
Allan, D.S.	F/O		4	
Allan, H.J.	LAC	511970	9	Also 36/37 as Cpl
Allan, J.D.	Sgt	11175	1,2	
Allan, W.J.	P/O		1	31 Squadron
Allen, C.H.	Sgt	227362	4	
Allen, C.E.H.	F/O		5	
Allen, F.L.	Cpl	361769	6	
Allen, H.H.		78916	1,3	
Allent, J.F.	LAC	12677	1	
Allen, V.E.	LAC	335268	4	Also took part in Waz 25 opns
Allen, W.C.	AC1	507821	6	N.R.I.V.

Name	Rank	No.	Clasp	Comments
Allen, W.E.	A/Cpl	10669	1	
Allerton, O.D.	F/O		6	
Allison, A.D.	LAO	334100	(3)	31 Squadron att. 20 Squadron
Allison, G.H.	F/O		1	Medal engraved 'G.Allison'
Alloway, H.Y	P/O		9	Also 36/37 as F/O. 28 Squadron
Alms, E.L.W.H.	F/O		5	
Amans, S.N	AC1	370196	6	
Amans, W.L.	AC1	505629	6	
Amner, A.J.	LAC	361844	6	
Ancell, H.B.	AM1	406407	1	
Anderson, A.M.	P/O	1 , (3)	Medal O.C.	
Anderson, C.	AC2	211277	2 , 3	
Anderson, C.E.	AC2	144528	1	
Anderson, C.T.	AC1	505601	6	
Anderson DFC AFC, D.F.	F/Lt		6	
Anderson, G.M.	F/Lt		6	
Anderson, G.McF	AC1	513555	9	Also 36/37 as LAC
Anderson, G.S.	Cpl	333747	8	
Anderson, J.M.	LAC	370890	6	
Anderson, W.	LAC	505707	6	
Anderson, W.D.	F/O		9	
Anderson, W.J.	AC2	281494	1 , 2 , 3	
Andrewartha, W.P.	LAC	359833	6	
Andrewartha, E.V.E.	F/O		5	
Andrews, E.J.	Cpl	335845	6	
Andrews, F.W.	Cpl	366008	6	
Andrews, H.	AC2	275068	1,3	52 Wing
Andrews, H.A.	AC1	346439	4	
Andrews, J.B.	LAC	94552	( )	Special List
Andrews, R.J.	LAC	354419	6	Also 37/39
Andrews, T.B.	LAC	347146	4	21/24 in lieu Waz 25
Andrews, W.	AC2	156549	1	
Andrews, W.A.	LAC	359068	6	
Angelo, G.G.	Cpl	343433	5	
Ankers DCM, B.	F/Lt		6	Also 36/37. Deceased
Annesley, W.R.B.	F/Lt		4	
Antelme, P.F.	Lt		1	
Anthony, A.	AM1	404269	1	
Anton, M.D.	AC1	343355	4	Also took part in Waz 25 Opns
Appleton	Sgt		1	
Appleton, W.	LAC	505085	6	
Arbuthnot, T. J.	F/Lt		9	
Archer, A.J.	Cpl	353898	6	
Archer CBE, J.O.	W/C		6	
Archer, L.T.	LAC	358722	6	
Ardley, E.L.	F/Lt		4	
Armstrong, G.E.	LAC	248263	(3)	99 Squadron
Armstrong, G.L.	LAC	590182	9	
Armstrong, I.	LAC	506935	6	
Armstrong, S.C.	AC1	239018	4	Also took part in Waz 25 Opns
Arney, F.	AC2	266031	1	
Arney, F.T.	AC2	260238	1	20 Squadron

Name	Rank	No.	Clasp	Comments
Arney, F.T.	AC2	341882	3	
Arnfield, S.J.	LAC	564115	9	Also 36/37
Arnold, F.A.	AC2	197023	1	
Arnold, L.C.P.	Cpl	508855	9	
Artis, H.A.	Cpl	343043	4	
Ash, A.O.	LAC	347765	4	
Ash, W.	LAC	507081	6	
Asher, R.	AC2	286393	1	
Ashford, J.H.	AC2	131287	1	
Ashley, C.H.	Sgt	561992	9	Also 36/37 60 Squadron
Ashmore, J.	Cpl	564323	9	Also 36/37 60 Squadron
Ashton, J.R.	AC2	507805	6	
Ashton, J.W.	AM2	18659	1	
Ashworth, H.R.	Cpl	14942	(3)	99 Squadron
Askew, J.	LAC	512678	9	
Astin, A.	LAC	84734	3	Entitled as AC1
Astin, J.R.	2/Lt		1	
Astley, N.	AC1	510967	9	
Aston, A.J.	Sgt	239138	6	
Atcherley, D.F.W.	F/Lt		8,9	
Atherley, J.E.M.	F/Lt		(3)	52 Wing. Adjudant
Atkins, J.C.	F/O		6	
Atkinson, B.F.	LAC	16776	1	31 Squadron
Atkinson, J.S.	Cpl	590084	9	
Atkinson, T.W.	Sgt	349828	9	
Atterbury, H.W.	LAC	342187	(5)	
Attfield, J.A.	LAC	355470	6	
Attree, A.E.	F/Sgt	38	5	
Attwell, E.	LAC	353538	6	
Austin, A.A.	F/Sgt	405317	6	
Austin, C.F.	AM2	31561	1	
Austin, F.C.	LAC	560505	9	Also 36/37
Austin, F.G.E.	AM2	105446	1	
Auton, J.R.	LAC	563515	9	36/37 as Cpl and 37/39
Auys, L.H.C.	Sgt	362487	6	
Avery, F.W.	LAC	370633	6	
Avery, L.H.	AC2	298578	3	
Avis, F.C.N.	LAC	363899	6	
Axhorn, J.	AM2	404268	1	
Axworthy, W.R.	1AC	507265	6	
Ayers, R.A.	LAC	561987	9	Deceased
Ayling, E.T.	AC1	327774	6	Deceased
Ayling, J.R.	F/O		7	
Ayres, C.A.	LAC	366001	9	
Babbage, G.S.	AC1	330578	4	21/24 in lieu Waz 25 Reduced to AC2 13.11.25
Bacon, G.F.	AC1	250057	(3)	
Bacon, J.L.W.	F/O		1	31 Squadron
Badsey, S.B.	Sgt	109253	1,3	
Baglin, B.A.	F/Sgt	204741	4	

Name	Rank	No.	Clasp	Comments
Bailey, G.	AC2	370143	6	
Bailey, H.	AC1	590498	9	Also 36/37 as LAC
Bailey, L.M.	S/L		5	
Bailey, S.	AM1	187361	1	
Bain, R.E.	F/Lt		6	
Bain, W.	LAC	292508	5	
Bainbridge, W.	LAC	359588	6	
Baker, A.G.L.	AC2	350263	5	
Baker, C.	LAC	364426	6	
Baker, E.	AC2	136893	1	
Baker, E.	LAC	238836	1	
Baker, E.I.	LAC	350254	4	
Baker, G.	F/Lt		3	Equipment Officer. 52 Wing
Baker, G.W.	LAC	361914	6	
Baker, G.W.W.	LAC	363343	6	
Baker, H.T.	LAC	361852	6	
Baker, J.E.	AC2	20116	1	
Baker, J.W.	F/Lt		5	
Baker, S.P.	Cpl	341147	8	
Baker, W.G.	AC2	157315	4	
Baldwin, A.C.	AC1	507661	6	
Baldwin, C.J.H.	AC2	508344	6	
Baldwin, E.P.	At2	512470	8,9	Also 36/37 as Cpl
Baldwin, E S.	Cpl	186185	6	
Baldwin, R.W.	Sgt	335589	8,9	Also 36/37
Balfe, F.	AC1	48346	(1)	5 Squadron
Balkwill, B.G.	AC1	78911	1	
Ball, A.E.	AC1	337854	6	
Ball, A.W.	AC1	515682	9	
Ball, C.D.	2/Lt		1,(3)	
Ball, F.S.	LAC	364496	9	Also 36/37
Ball, J.W.	LAC	365182	8	
Ball, W.C.	AC2	244133	1	
Ballantyne, J.	LAC	365162	6	
Ballard, L.	Cpl	361863	6	
Balsdon, C.	Cpl	347697	4	A/Sgt. 21/24 in lieu Waz 25.
Balshaw, J.G.	Cpl	361930	6	
Balsom, S.J.H.	LAC	366051	9	Also 36/37
Banks, R.P.	AC2	62000	1	
Bannett, C.D.	AM2	39970	1	
Banwell, G.E.	Sgt	340755	9	
Barber, C.P.	F/Lt		5	
Barber, N.	AC1	561472	9	Also 36/37
Barclay, D.	F/O		6	
Barclay, R.B.	LAC	366056	9	Also 36/37
Barclay, R.A.C.	F/O		9	
Bardsley, H.	LAC	356521	6	
Barker, A.T.	AC1	370324	6	
Barker, C. G.	F/O		1	31 Squadron
Barker, C.L.O.	LAC	364430	6	
Barker, G.K.	AC1	507218	6,8	
Barker, G.T.	LAC	563835	9	

Name	Rank	No.	Clasp	Comments
Barker, H.G.	LAC	512037	9	
Barker, J.W.	AC2	509553	6	A/Cpl
Barker, W.H.	AC2	108867	3	
Barley, A.E.	LAC	330061	5	
Barlow, A.A.E.	AC1	557696	6	
Barlow, T.M.	AC2	293412	3	
Barnard, F.H.	AC1	345352	4	
Baron, H.	Cpl	22701	1	
Barr, C.G.W.	LAC	370867	6	
Barras, J.C	Sgt	333729	(3)	20 Squadron
Barrett, A.W.B.	LAC	365652	6	
Barrett, H.E.	LAC	364491	6	
Barrett, H.H.	AC1	342298	5	
Barritt, R.	Cpl	404258	1	
Barron, A.	AM2	111099	1	
Barrow, L.V.G.	F/O		6	
Bartell, M.	LAC	590285	9	Also 36/37 as Cpl
Bartlett, L.E.	LAC	511993	9	Also 36/37 as Cpl
Bartlett, R.G.	Cpl	327584	5	
Bartlett, R.J.O.	F/O		6 , 8	
Bartlett, R.W.	LAC	511278	9	Also 36/37
Bartlett, W.L.	Sgt	5733	1	
Barton, A.	LAC	10092	1	
Barton, E.	ACl	345412	4	
Barton, J.S.	ACl	370393	6	
Barton, M.I.	AC1	56396	(3)	99 Squadron
Barton, J.R.F.	W/C		(3) , 4	28 Squadron
Barton, W.	LAC	363803	6	
Baseden, M.W.	F/O		(3)	
Basham, H.H.	Cpl	326018	4	
Bashford, J.D.	Cpl	327353	4	
Bashford, R.C.	AC1	S06786	6 , 8	
Bass, E.F.	AC2	346480	4	
Batchelor, C.T.	LAC	359298	6	
Bates, A.S.	LAC	370726	6	
Bates, C.W.	LAC	362782	6	
Bates, G.	LAC	355701	6	
Bates, J.D.	AC1	515684	9	
Batten, A.H.	LAC	363799	6	
Battersby, P	AM2	35885	1	
Battle, H.F.V.	F/Lt		9	
Baughan, E.W.	AC1	336516	(3)	20 Squadron
Baughan, R.A.	SM1	333	4	
Baxter, C.A.	Sgt	327220	9	
Baxter, E.	A/Cpl	104669	(1)	20 Squadron
Bayford, F.C.	Cpl	96294	(3)	97 Squadron
Bayman, C.A.	LAC	350093	4	21/24 in lieu Waz 25
Bayne, D.W.	F/O		6 , 8 , 9	
Beachman, H.	AC2	37906	1	14 Squadron
Beagarie, G.D.	LAC	561456	9	Also 36/37 as Cpl
Beal, W.	LAC	513625	9	Also 36/37 and 37/39
Beale, T.E.	LAC	560523	9	

Name	Rank	No.	Clasp	Comments
Beales, H.G.	LAC	363930	6	
Beaman, W.R.	F/O		6	
Beard, F.J.R.	AC1	513027	9	
Beardsall, F.E.	AC1	158844	(1),(2)	
Bearman, S.F.	F/Sgt	366022	9	Also 36/37 as LAC
Beasley, H.	AM2	25450	1	
Beasley, T.C.	LAC	354707	6	
Beasley, W.C.	AM2	78927	1	
Beatson, B.F.	S/L		4	Previous service with IMS
Beaumont, C.G.	AC1	334073	4	28 Sguadron (1932)
Beaumont, F.B.	AC2	239583	1,3	99 Sguadron (31.3.25)
Beaumont, J.F.	Cpl	349604	6	
Beaven, E.A.	LAC	353803	6	
Beavan, L.H.	AC1	326487	5	
Beazer, J.F.D.	AC2	246844	1,3	
Beck, H.	F/Sgt	403560	6	
Beck, L.W.	P/O		1	5 Squadron. Medal engraved 'L. Beck'. Issued 27.4.23
Beck, R.W.	LAC	363366	6	
Becker, B.H.	F/Lt		9	Also 36/37. 5 Squadron
Beddow, W.P.	Cpl	358181	6	
Bedford, C,C		104683	1,2,3	31 Squadron.
Beech, F.C.		138885	1	31 Squadron
Beeching, R.H.	F/Sgt	365668	9	Also 36/37 and 37/39
Beer, A.A.J.	F/Sgt		6	
Begg, J.A.B.	F/Lt		9	Also 36/37 and 37/39 (N of K)
Beith, R.D.	AC1	511592	9	
Belcher, R.W.	LAC	28439	1	52 Wing
Belfe, L.	AC1		1	
Belford, J.C.	F/O		4	
Bell, A.	Cpl	341373	4	
Bell, A.A.	Sgt	15540	4	
Bell, J.F.	LAC	358774	6	
Bell, L.	LAC	38269	( )	5 Squadron.
Bell, R.L.	F/Sgt	201227	7	
Bell, S.F.	Cpl	20349	1	
Bell, W.	Sgt	7405	1	
Bellamy, E.	AC1	343845	5	
Bell-Irving, L.H.	F/O		3	
Bellis, J.A.	AC1	331813	(3)	Further details requested
Belsham, C.N.	LAC	366023	6	
Bennett, C.H,	LAC	326247	(5)	
Bennett, E.G.	LAC	363825	6	
Bennett, J.M.	Sgt	275	1	
Bennett, K.A.	LAC	561036	9	Also 36/37
Bennett, R.	Sgt	352392	9	Also 36/37
Bennett, R.	LAC	590108	9	
Bennett, R.S.	Sgt	2588	1	
Bennett, W.T.	LAC	362783	6	
Benney, H.N.	LAC	365167	6	
Bennion, G.F.	AC2	511806	6	
Benson, A.	AC2	143492	(3)	

Name	Rank	No.	Clasp	Comments
Benson, J.M.	AM2	76480	1	
Bentley, G.E.	LAC	333799	6	
Bernard, A.H.	LAC	516306	9	A1 at 36/37
Bernard, J.L.	Lt		1	114 Squadron
Berry, T.J.	LAC	560533	9	Also 36/37
Berry, W.S.	Cpl	157978	6,9	
Berryman, P.E.	F/Lt		6	
Bertram, I.A.	F/Lt		6	
Bessey, C.L.	LAC	364498	6	
Best, C.L.	LAC	364416	6	
Best, W.	F/O		4,5	
Bestwick, T.R.	AC1	340219	3	20 Squadron
Beswick, J.W.	LAC	39849	1	
Betteridge, S.A	LAC	562070	9	
Betty, S.G.	F/Sgt	362786	9	Also 36/37 60 Squadron
Bevan, E.J.	LAC	505479	6	
Bewsey, D.	SM	1499	1	
Bexon, J.	LAC	331633	6	
Bickenson, J.J.S.	AC1	363881	6	
Bickley, L.J.	Cpl	364510	9	
Bicknell, A.J.	LAC	327219	4	
Biddall, A.	AC1	511471	9	
Biddle, R.	AC1	337924	(3)	22 Squadron
Bidmead, A.	AC2	33034	4	28 Squadron (1932)
Biggar, A.J.	F/O		9	
Biggins	LAC	220721	1	
Bignell, H.J.	Cpl	363856	6	
Bilbo, E.W.	LAC	365183	6	
Billing, W.R.	AC1	326437	4	
Billinger, C.H.W.	Cpl	363533	6	
Billson, R.A.	LAC	363960	8	
Bilner, C.A.	AC1	342002	4	
Bilney, C.N.H.	F/Lt		6	
Bindon, I.V.	Cpl	50437	1	
Bingham, W.P.	O/O		1	
Binns, H.D.	LAC	363540	6	
Binstead, W.V.	LAC	365163	6	
Birch, J.A.	AC1	511446	9	
Birch, S.W.	SM2	4174	6	
Bircham, C.G.	F/Sgt	9360	( )	Not entitled to Afg 19/21
Bird, A.E.	LAC	327590	4	
Bird, C.H.	F/O		3	20 Squadron
Bird, G.	Cpl	335359	6	
Birtwistle, R.H.	LAC	505993	6	
Bishop, A.G.	F/Lt		5	
Bishop, G.J.	Cpl	359173	6	
Bishop, H.L.	LAC	359043	6	
Bishop, P.E.	F/Lt		1,2,3	
Bishop, T.E.	LAC	365147	6	
Blackburn, G.F.	F/O		4	
Blackden, V.Q.	F/O		6	
Blackman, F.	F/Sgt		1	5 Squadron

Name	Rank	No.	Clasp	Comments
Blackmore, R.M.	AC2	247582	(3)	
Blackmore, W.G.E.	F/Sgt	3125	(3)	20 Squadron
Blackwood, G.	AC2	329347	1,2,3	
Blackwood, J.J.	AC2	507683	6,8	
Blain, W.N.	F/O		6	
Blake, A.R.	LAC	507236	9	
Blake, C.H.	AM3	187866	1	
Blake, E.A.	F/O		4	
Blake, G.L.	F/O		2,3	
Blake, F.G.	LAC	359309	6	
Blake, R.J.	LAC	364537	6	
Blake, T.A.	Cpl	515317	9	Also 36/37 and 37/39
Blake, W.R.	AC2	239802	(3)	99 Squadron
Blakeman, L.H.	AC1	564541	9	Also 36/37
Blanchard, F.	AC1	341016	4	
Blanche, H.E.	A/Sgt	240535	5	Deceased
Blenkarn, J.E.	AC1	507926	6	
Blenkinsope, G.W.	LAC	359839	6	
Blick, J.F.	F/O		1,2,3	20 Squadron.
Block, G.T.	Sgt	342563	6	
Bloomer, F.	AC1	506458	6	
Blowers, C.W.	LAC	505452	6	
Bly, H.N.	Cpl	356198	6	
Blyth, A.E.	LAC	341138	4	
Board, A.C.	AM2	405830	1	
Board, E.	LAC	356988	6	
Bodley, J.E.	LAC	507850	1	Appn for 1930/32 disallowed
Body, G.A.	P/O		1	
Body, J.H.	F/O		(1)	31 Squadron. Medal O.C.
Bodycomb, E.A.	AC1	124S71	1	
Boggis, W.G.	AC1	330901	5	
Boland, J.	LAC	365154	6	
Bolsover, J.R	LAC	359707	6	
Bolton, A.W.	Cpl	513276	9	Also 36/37
Bolton, C.V.D.	LAC	510655	8,9	
Bolton, E.R.	Cpl	348467	6	
Bond, A.G.	F/Lt		4	
Bone, A.L.	AC2	327982	4	
Bone, G.	F/Sgt	335457	9	Also 36/37
Bonehill, F.C.	AC2	332142	(3)	20 Squadron
Boniface, H.A.	F/O		5	
Boniface, J.	LAC	327892	5	
Bonney, C.T.J.	LAC	363774	6	
Boon, J.E.	LAC	326446	4	
Boon, R.C.	LAC	361847	6	
Boorer, H.J.F.	AC1	56493	3	
Booth, C.B.	LAC	561043	9	Also 36/37
Booth, F.	AC2	118461	(3)	27 Squadron. N.R.I.V.
Booth, J.G.	Cpl	355316	6	
Boothroyd, C.G.	P/O		1	20 Squadron.
Boots, W.T.	Cpl	344779	9	Also 36/37 and 37/39
Boreham, W.E.	Sgt	200414	1	

Name	Rank	No.	Clasp	Comments
Boret, J.A.	F/O		4	28 Squadron (1932)
Borley, C.A.	AC1	370586	6	
Borradaile, F.M.C.	AC1	506960	6	
Bostock, H.B.	LAC	365658	9	
Boston, F.	F/O		5	
Bothwick, J.F.	LAC	513093	9	Also 36/37
Bottomley, N.H.	G/C		9	Also 36/37. Air Vice Marshall
Bottoms, L. S.	LAC	348618	4	Also 36/37 as Sgt
Boucher, A.J.	LAC	357978	6	
Boucher, H.A.	AC1	406540	1	31 Squadron
Bouchier, C.A.	F/O		3	
Bounden, S.	LAC	157808	2,3	
Boundy, R.	AC1	59853	( )	20 Squadron.
Bourne, W.H.A.	Cpl	364539	(7)	
Bouser, H.	AC1	109798	1	31 Squadron
Bowdidge, J.S.	SM1	302794	6	
Bowditch, A.L.	AC2	406184	1	
Bowen, D.I	Cpl	333361	6	
Bowen, T.I.	AC2	506742	6,8	
Bower, L.W.C.	P/O		6	
Bowers, F.S.	Cpl	334935	(3)	99 Squadron
Bowers, W.J.	Cpl	247225	5	
Bowes, G.B.	Cpl	341816	5	
Bowes-Lyon, H.M.	P/O		6	
Bowker, W.A.	AC2	269116	1	
Bowley, L.	AC2	334195	4	
Bowman, F.L.	Cpl	340824	(3),6	52 Wing
Bowman, W.	Cpl	313630	9	
Bown, A.J.	Sgt	4429	1	
Bowring, A.	Cpl	336127	(3)	31 Squadron attd 20 Squadron
Boyd, L.C.	F/Lt		1	31 Squadron
Boyd, W.	AC1	136198	( )	5 Squadron.
Boyd, W.L.	AC1	191022	1,2,3	
Boyes, J.W.	AM2	148719	1	
Boyes, W.H.	AC1	24076S	2,3	20 squadron
Boyle, F.	LAC	362256	6	
Boyles, F.W.	Cpl	355033	6,8	
Boylin, F.L.	LAC	355452	6	
Bracey, H.J.	AC2	243608	3	20 Squadron
Bradbury, J.	F/O		4	21/24 in lieu Waz SS
Bradford, A.	AC2	156103	2,3	20 Squadron
Bradley, A.	LAC	404193	1	
Bradley, A.E.	LAC	509S17	9	Also 36/37 as Cpl
Bradley, E.W.	LAC	328656	(5)	
Bradley, H.	LAC	332119	3	20 Squadron
Bradley, J.	Sgt	342491	9	Also 36/37
Bradshaw, J.	F/Sgt	250247	4	
Bragg, W.J.A.	Sgt	349705	6	
Bragg, W.S.	AM1	18938	1	
Bramble, L.E.	Cpl	358026	6	
Brand, A.S.	AM1	3961	1	
Brandon, E.J.	Cpl	560557	6,9	

Name	Rank	No.	Clasp	Comments
Brandon, T.G.	LAC	509248	9	
Brannan, A.E.F.	LAC	359133	6	
Braund, E.W.	Sgt	157159	5	
Bray, R.	AM2	26832	1	
Braye, I.W.	F/O		9	Also 36/37 and 37/39.MID.60
Squadron				
Brazier, J.S.	Cpl	343455	6	
Breakell, T.R.	P/O		4	
Breed, A.E.	LAC	359540	6	
Brerton, G.W.	Sgt	344714	9	Deceased
Bretherton, N.C.	F/O		5	Medal broken, new medal issued
Brewerton, C.F.	F/O		3,(4)	20 Squadron. Killed 1927.
Brewin, F.C.	LAC	345742	4	
Briant, C.H.	LAC	240872	4	21/24 in lieu Waz 25
Briant, F.	AM1	1413	1	
Briard, J.	Lt		1	1/35 Sikhs and 31 Squadron
Bricknell, E.E.	Lt		3	
Bridge, B.H.	Cpl	365144	6	
Bridge, W.A.	LAC	512001	9	Also 36/37
Bridsen, J.M.	AC1	64599	1	5 Squadron
Brie, R.A.C.	O/O		2,3	
Brierly, G.H.	AC2	344985	4	Also 36/37 & 37/39
Briggs, E.N.	LAC	562455	9	
Briggs, H.	AM2	107301	1	
Briggs, J.	Cpl	350770	6	
Bright, J.L.	LAC	363346	6	
Brighton, W.F.	LAC	349715	5	
Brimfield, L.	Sgt	359168	6	
Brinsley, A.H.	LAC	511440	9	
Briscoe, A.	F/Lt		3	20 Squadron
Bristow, L.W.	LAC	45081	1	
Bristow, L.W.	LAC	145981	1	114 Squadron
Brittain, H.F.	LAC	508577	6,9	
Britter, W.B.	LAC	337454	3	52 Wing
Britton, A.S.	Cpl	62035	4	
Britton, C.F.G.	LAC	505551	6	
Britton, E.A.	AM1	11351	1	
Britton, E.A.C.	F/O		1,2,3,6	Medal O.C.
Broadhurst, H.	F/O		6	
Broadis, A.J.	A/Sgt	3249	1	
Brock DSO, H.Le M.	G/C		6	
Brock, R.W.	AC2	507613	6	
Brockerton, G.E.	Cpl	560520	9	Also 36/37
Bromham, F.J.	AC2	343700	4	21/24 in lieu Waz 25
Bromley, E.	AC1	343662	4	Also took part in Waz 25 Opns
Brook, W.A.D.	F/O		4	21/24 in lieu Waz 25
Brooke, A.F.	F/Lt		3,6	
Brooke, F.	LAC	506998	8	
Brooker, T.	AC2	275918	1	
Brookman, H.G.	F/Lt		9	Deceased
Brooks, A.J.	LAC	364473	6	
Brooks B.J.	Cpl	364492	9	

Name	Rank	No.	Clasp	Comments
Brooks, F.L.	AC1	513681	9	
Brooks, W.C.	AC1	332601	3	20 Squadron
Brooksbank, W.	AC2	64320	1	
Broomhead, R.W.	LAC	344462	5	
Brotherhood, W.R,	F/O		9	37 Squadron
Brough, H.J.	LAC	357809	6	
Broughton, C.	F/O		9	37 Squadron
Brown, A.J.	AC1	512244	9	
Brown, A.K.	LAC	356607	6	
Brown, D.	LAC	341904	4	21/24 in lieu Waz 25
Brown, E.A.	Sgt	309179	9	Also 36/37. Deceased
Brown, F.J.	AC2	244723	2 , 3	
Brown, F.R.	AC1	90745	4	
Brown, F.W.	LAC	564057	9	Also 36/37
Brown, G.D.	LAC	354593	6	
Brown, G.F.	LAC	341922	4	
Brown, G.H.	LAC	342788	4	
Brown, G.H.	LAC	563520	9	Also 36/37
Brown, G.W.	LAC	364513	9	
Brown, H.J.	F/O		4	
Brown, H.S.	LAC	358642	6	
Brown, J.E.	F/Sgt	87076	9	
Brown, J.F.	Cpl	358701	6	
Brown, J.H.	F/O		6	
Brown, J.T.	SM1	202020	6	
Brown, K.H.	F/O		1	31 Squadron
Brown DSC AFC, L.O	S/L		6	
Brown, L.S.T.	F/O		6	
Brown, P.	F/Sgt	241914	9	
Brown, R.	AM1	50913	1	
Brown, R.F.	LAC	363547	6	
Brown, R.F.	LAC	510712	9	Also 36/37
Brown, S.G.	AC2	158167	1	
Brown, S.G.	Cpl	362485	9	Also 36/37. Forfeited
Brown, T.	Sgt	12685	1	
Brown, T.H.	AC1	248024	3	
Brown, W.	LAC	358739	6	
Brown, W.G.	AC2	81791	2 , 3	
Brown, W.J.L.	Cpl	159387	4	21/24 in lieu Waz 25
Browne, A.C.B.	Sgt	250447	9	
Browne, W.D.	LAC	362244	6	
Brownlee, J.	F/Sgt	313644	4	
Brownridge, W.	T/Sgt	218844	( )	20 Squadron.
Bruce, W.R.	S/L		1	No.1 Indian Wing
Bruford, M.B.	LAC	347026	(5)	N.R.V.I.
Brummell, A.W.	AC1	359947	6	
Brundle, E.J.	LAC	366031	9	
Brundle, E.W.	LAC	362280	6	
Bryant, C.	Cpl	510304	9	
Bryant, E.G.	LAC	515352	9	Also 36/37
Bryant, G.E.	Sgt	345657	6	
Bryer OBE AFC, G.M.	F/Lt		6	

Name	Rank	No.	Clasp	Comments
Bryson, O.C.	S/L		6	
Bubb, E.A.	LAC	341751	4,6	
Bubb, H.J.	LAC	358747	6	
Bubb, H.N.	Cpl	353632	6	
Buchanan, J.W.	F/Lt		9	Also 36/37
Buckland, A.	AM2	406541	1	
Buckle, W.E.L.	Cpl	327862	4	
Bucklow, G.	LAC	365161	6	
Buckman, D.A.	AC1	507824	6	
Bucknall, H.H.	Sgt	63042	6	
Bucknole, H.V.	LAC	361834	6	
Budd, G.	AM2	17905	1	
Budden, H.G.D.	AC1	512808	9	
Budworth, L.E.	AC1	359468	6	
Buesnel, H.E.	Sgt	12592	1	Deceased
Bugg, E.E.A.	Sgt	1956	6	
Buggy, C.K.	Cpl	359489	6	
Bugh, L.P.	AC2	241107	1,3	Medal O.C.
Bullen, F.W.	LAC	565670	6,8,9	
Buller, W.	AN2	240710	1	
Bulley, E.T.M.	AC1	358538	6	
Bunch, J.W.R.	LAC	356009	6	Medal and clasp Waz 21/24 from Army service
Bunyan, M.D.	LAC	316771	4	21/24 in lieu Waz 25
Burbridge, F.C.	Cpl	213098	3	20 Squadron
Burden, W.S.B.	LAC	359576	6	
Burdett, G.F.	SM2	87728	6	
Burford, A.O.	LAC	347901	5	Approved in rank of Cpl
Burford, H.J.	AC1	2742	( )	5 Squadron. N.R.V.I.
Burge, H.	LAC	358705	6	
Burge, R.C.	LAC	363796	8	
Burgess, A.J.	LAC	344845	4	Deceased
Burgess, F.	AM2	300488	1	
Burgess, J.	LAC	506824	6	
Burgess, W.C.H.K.	LAC	355424	6	A/Cpl
Burke, N.	F/O		1	31 Squadron
Burleigh, A.H.	F/Sgt	242080	9	
Burley, A.R.	AC1	344350	5	
Burley, J.W.	LAC	346262	5	
Burman, F.M.	AC1	328540	5	
Burnett, J.R.	AC1	511963	9	
Burnett, K.D.	LAC	562642	9	Also 36/37
Burnley, D.C.	P/O		5	
Burns, E.A.	Cpl	328222	9	
Burns, W.	AC2		1	
Burnside, C.S.	LAC	361893	6	
Burridge, W.A.	Cpl	356893	9	
Burrows, B.E.	Sgt	2993	( )	31 squadron
Burrows, C.	AC2	131094	3	
Burrows, F.O.	F/Sgt	326698	( )	N.R.V.I.
Burrows, J.	F/Sgt	222492	9	Also 36/37
Burrows, S.W.	AC1	327394	5	

Name	Rank	No.	Clasp	Comments
Burt, F.G.B.	AC1	513533	9	Also 36/37 as LAC
Burtenshaw, H. J.	LAC	513460	9	Also 36/37
Burton, A.P.	AC2	507544	6	
Burton, E.	F/Lt		4	
Burton, F.	LAC	560021	9	
Burton, J.	LAC	342454	4	
Burton, W.C.	Sgt	240296	6	
Burton, W.C.E.	LAC	354894	6	
Bury, W.T.	AM2	748591	1	
Bush, F.L.	LAC	343050	4	
Busk, C.W.	F/Lt		5	
Bussell, E.I.	F/Lt		1,2,3,4	
Butcher, A.H.	AM2	302537	1	
Butcher, H.T.	LAC	514875	9	
Butcher, J.E.	LAC	356545	6	
Butcher, L.C.	Cpl	361856	6	
Butcher, W.G.H.	F/Sgt	314957	4	21/24 in lieu Waz 25
Butler, H.S.	LAC	344417	5	
Butler, J.	AC2	345716	5	Waz 25 in lieu 21/25
Butler, J H.	F/O		1,2,3	99 Squadron. Also 36/37
Butler, J.W.	AC1	509852	6	
Butler, P.	Cpl	309740	6	
Butler, P.B.H.	F/Lt		9	Also 36/37 & 37/39. 27 Squad.
Butler, R.H.	AC1	363867	6	
Butler, S.	AM1	132271	1	
Butler, W.D.	F/O		6	
Butler, W.J.	F/Lt		1	31 Squadron
Butt, L.A.K.	F/Lt		(3)	20 Squadron
Button, E.	AM2	148486	1	
Byng, F.	AC2	275894	1	
Caddy, W.	LAC	370492	6	
Cadman, H.	SM2	204479	6	
Cafferkey, J.P.	F/O		1,3	
Cain, J,C.	LAC	352956	6	
Cairns, A.H.	AC2	507733	6,8,9	
Cairns, G.	AC1	347556	5	
Cairns, J.H.	AC1	334419	3	52 Wing
Cake, P.J.R.	AC1	263348	(3)	97 Squadron
Calder, D.Mc.D.	P/O		1	20 Squadron
Calder, G.J.	AC1	561494	9	
Callagan, J.L.	LAC	510424	8,9	
Callicott, F.G.	AC1	156111	4	21/24 in lieu Waz 25
Calpin, H.	AC1	505927	6	
Calvey, H.C.	F/Lt		4	
Cameron, D.	AC1	514304	9	Also 36/37
Campbell, D.A.	AC2	340902	4	
Campbell, D.H.	LAC	362865	6	
Campbell, F.	LAC	512308	9	
Campbell, G.E.	Sgt	90741	5	F/O (1923)
Campbell, H.	AC1	123200	1	31 Squadron

Name	Rank	No.	Clasp	Comments
Campbell, J.	W/o	337712	9	
Campbell, L.	LAC	512156	9	
Campbell, M.	AC1	330399	5	
Campbell, W.McC.	AC2	509497	6	
Campion, F.	LAC	359771	6	
Cann, W.H.	LAC	343379	(5)	
Canniford, W.H.	Sgt	355178	9	
Canning, P.F.	P/O		6	
Cannock, W.M.E.	W/O	254131	9	
Cantor, N.R,	AC2	343131	4	
Capol, A.J.	S/L		5	Waz 25 in lieu 21/24
Capp, L.W.	LAC	561091	9	A/Cpl
Capps, H.V.	AC1	326427	5	Waz 25 in lieu 21/24
Carbery, D.H.M.	F/Lt		1 , 3	31 Squadron. Medal O.C.
Card, A.R.	Cpl	329052	4	
Card, L.T.	Sgt	362296	6	
Carden, L.C.	LAC	348809	4	
Careless, E.A.	AC1	337431	(3)	20 Squadron
Carew, R.P.H.	F/O		9	
Carey, V.C.	LAC	506729	6	
Carley, F.	AC1	157479	3	99 Squadron
Carlielle, A.J.	Cpl		(4)	Cpl RASC served in RAF
Carpenter, A.T.G.	LAC	361948	6	
Carpenter, C.N.	F/O		8 , 9	
Carroll, G.N.	F/O		5	
Carter, A.S.	AC1	511725	9	
Carter, B.C.B.	AC1	341717	(5)	
Carter, F.C.	LAC	362283	6	
Carter, F.G.	P/O		1	Medal O.C.
Carter, F.V.	LAC	364542	6	
Carter, N.	F/O		4,9	21/24 in lieu Waz 25
Carter, R.A.C.	F/O		9	
Carter, R.J.	AC1	326837	4	
Cartwright, E.	AM1	1538	1	
Cartwright, R.	LAC	36739	( )	
Carwood, H.E.	AC1		1	
Cary, B.M.	F/O		6	
Casey, R. F .	F/O		4	
Cassey, T.	SM1	301755	4	21/24 in lieu Waz 25
Cassidy, M.J.	F/Sgt	26233	(3)	20 Squadron
Castle, R.E.	LAC	350326	5	
Caswell, E.	AC2	249199	( )	5 Squadron
Cathey, J.R.	LAC	353719	6	
Catton, W.A.	AC1	357993	(6)	
Cattrell, J.McG.	Cpl	349587	6	
Causon, J.J.	Cpl	561501	9	Also 36/37
Causon, W.E.	LAC	365703	6	
Cauthery, R.P.	F/O		6	
Cave, C.T.	LAC	330090	(5)	N.R.V.I.
Cave, R.S.	AC2	340783	(3)	20 Squadron
Chacksfield, B.A.	F/Lt		9	Also 36/37
Chadwell, C.A.	Cpl	6120	1	

Name	Rank	No.	Clasp	Comments
Chadwick, J.L.	AC1	336090	4	
Chadwick, R.	LAC	343192	4	21/24 in lieu Waz 25
Chadwick, S.	AM2	77747	1	
Chaffey, F.G.	AC1	227971	3	
Chalk, S.G.	W/O	8766	9	Entitled as F/Sgt
Chamberlain, G.P.	F/O		6	
Chamberlayne, P.R.T.J.M.I.C.	F/Lt		5	
Chambers, A.W.C.	AC2	118698	1	
Chambers, C.A.	LAC	327640	(5)	N.R.V.I.
Chambers, F.A.	Cpl	199395	4	21/24 in lieu Waz 25
Chambers, I.J.	AC2	167756	1	20 Squadron
Chambers, J.E.	Cpl	561085	9	Also 36/37
Champion, H.	Cpl	94116	(3)	
Champion De Crespigny MC DFC, H.V.	F/Lt		2,3,4,6	97 Squadron
Champken, A.L.T.	LAC	357614	6	
Chandler, C.H.	SM1	416	6	Not entitled to Mohmand 1933
Chandler, G.W.	F/Sgt	381	1	
Chandler, J.A.	LAC	514437	8,9	Also 36/37
Chapell, H.J.	LAC	363555	(6)	Deceased
Chaplin, E.A.	LAC	510582	9	
Chapman, C.H.	LAC	359922	6	
Chapman, J.G.	LAC	341078	4	
Chapman, J.H.	AC2	116187	2 , 3	
Chappell, L.E.	Cpl	335151	5	Entitled as AC1
Chard, R.	AM2	48894	1	
Charles, A.G.	LAC	359504	6	
Charles, G.W.	AC1	403160	6	
Charlesworth, C. L.	LAC	34S655	4	Also took part in Waz 25 Ops
Charlesworth, E. R.	LAC	359557	6	Deceased
Charlegworth, P.A.V.	LAC	345759	4	
Charleton, C.	LAC	370168	6	
Charleton, D.	LAC	370500	6	
Charleton, F R	AC2	149954	3	99 Squadron
Chase, A.G.	AC1	340544	4	
Chase, J.J.	LAC	334135	3	20 Squadron
Chase, W.A.	F/O		5	27 Squadron
Chasemore, F.E.	LAC	343555	4	21/24 in lieu Waz 25
Chatt, A.	LAC	370075	6	
Chatterton, H.A.	A/Sgt	562047	9	Also 36/37
Chaytor, W.L.	LAC	511111	9	
Cheeseman, E.F.	AC2	245662	1	
Cheeseman, V.	LAC	4028	(I)	20 Squadron
Cheeseman, W.J.	AC1	512191	9	
Chenery, H.J.	LAC	561503	9	Also 36/37 as A/Sgt
Chesney, C.W.	LAC	560067	9	Also 36/37 as Cpl
Chick, J.S.	F/O		3,4	97 Squadron
Chidgey, E.S.M.	Sgt	348526	9	Also 36/37
Chidlow, E.	LAC	332807	3	20 Squadron
Chilcott, A.E.	Cpl	340586	6	
Childs, A.	AC1	332174	3	20 Squadron

Name	Rank	No.	Clasp	Comments
Childs, S.L.	LAC	362468	6	
Chilton, J.W.	AC1	330081	4	
Chipp, R.E.	AC1	363378	6	
Chittenden, O. I.	A/Cpl	358932	6	
Chown, J	AC1	330427	4	
Christian, C.W.	Sgt	11038	3	
Church, R.	LAC	366063	9	
Chuter, E.J.	Sgt	5768	1	
Chuter, F.D.	LAC	364451	6	
Clark, A.	AC1	281087	3	97 Squadron. Dup medal issued
Clark, A.M.	AC1	506039	6	
Clark, A.R.	Sgt	1015	4	
Clark, B.C.	AM2	12467	1	
Clark, D.	Cpl	10743	1	
Clark, L.W.	LAC	362878	6	
Clark, S.J.	LAC	364557	9	
Clark, W.F.	LAC	358575	6	
Clark, W.H.	LAC	505980	6	
Clark, W.J.	AC1	340839	3	97 Squadron
Clarke, A.J.	LAC	342525	5	
Clarke, A. R.	A/Sgt	315032	6	
Clarke, E.L.	Sgt	366200	9	Also 36/37
Clarke, G.	LAC	365503	6	
Clarke, H.	LAC	512054	9	Also 36/37 as Cpl
Clarke, H.A.	LAC	242063	4	
Clarke, F.F.	AC2	248332	1	
Clarke, J.H.	LAC	345950	(5)	N.R.V.I.
Clarke, J.J.	LAC	361853	6	
Clarke, O.C.	Sgt	330150	6	Deceased
Clarke, S.A.	Cpl	356808	6	
Clary, W.	SM2	314884	4	
Clay, F.	SM1	16046	4	
Clayson, A.E.	Sgt	8246	1,3	99 Squadron
Clayton, L.	LAC	341819	4	
Clayton-Daubeny H.F.	P/O		6	
Clegg, J.B.	AC1	35686	1	5 Squadron
Clements, F.	AM2	20793	1	
Clements, F.G.	AC2	248606	1	
Clements, T.A.	Cpl	354380	6	
Clements, W.G.	AC1	357217	6	
Clemmow, A.M.	LAC	357035	6	
Clemson, R.M.	AC1	515930	9	
Clifford, E.J.	AC1	327949	4	21/24 in lieu Waz 25
Clifton, F.	AC1	276675	3	20 Squadron
Cloete, D.	Major		1	114 Squadron
Clouder, F.J.	Sgt	I991	4	
Clow, C.J.	AC2	250538	2	
Clowes, J.E.P.	AC2	63604	1,3	
Clydesdale , W. R.	LAC	358092	6	
Clyne, M.F.W.	LAC	358671	6	
Coates, E.	LAC	348132	4	
Coatsworth, S.	AM2	13770	1	

Name	Rank	No.	Clasp	Comments
Cobbold, F.J.	LAC	347465	5	
Cockbain, H.P.	LAC	359888	6	
Cockell, G.F.	Cpl	248272	(3),6	99 Squadron
Cocker, J.W.	Cpl	348100	6	
Cockerline, A.L.	LAC	248608	3,4	97 Squadron.
Cockle, D.F.E.	LAC	365381	9	
Cockram, S.W.	Sgt	1344	5	
Cocks, J.J.C.	F/O		5	Deceased
Codrington, S.A.	LAC	505611	6	
Cogram, W.C.	AC1	560059	9	
Cogswell, F.C.	F/O		1,3	Medal O.C.
Cohu, J.M.	F/Lt		9	
Colam, E.E.F.	F/O		5	
Colbert, R.J.	LAC	514552	9	Also 36/37 as Cpl
Colbran, A.G.	AC1	340853	4	
Colder, C.E.A.	Sgt	248819	4	
Coldwell, R.S.	AM2	11036	1	
Cole, B.	LAC	370927	6	
Cole, F.C.	F/Lt		9	
Cole, S.H.	F/Sgt	213279	1	
Cole, W.F.L.	LAC	366072	9	
Colebrook, M.	LAC	509270	9	Also 36/37
Coleman, W.D.	SM1	10982	6	
Coles, C.J.	LAC	335888	6	Deceased
Coles, W.B.	AC1	247408	3	20 Squadron
Coley, E.J.	LAC	338594	(5)	N.R.V.I.
Collett, A.V.	SM2	289	6	
Colley, O.	Sgt	333019	6	
Collick, A.A.	Cpl	354533	6	
Collier, G.E.	Sgt	336016	6	
Collier, K.D.G.	F/Lt		4	
Colling, R.H.	AC1	515639	9	Also 36/37 and 37/39
Collings, R.W.P.	F/O		6	
Collingwood, C.J.	F/Lt		6	
Collins, A.B.	F/Sgt	4383	6,8	
Collins, F.	Sgt	5763	6	
Collins, F.	LAC	159596	5	
Collins, F.W.	LAC	330315	5	
Collins, H.M.B.	F/O		6	
Collins, J.	AC1	306315	6	
Collins, L.R.	Cpl	352862	8,9	
Collins, R.A.	Cpl	353490	9	Also 36/37 as Sgt
Collins, W.E.	LAC	364796	6	
Collis, E.J.	AC1	406340	1	114 and 31 Squadrons
Colson, V.S.	Cpl	342163	9	Also 36/37 as A/Sgt
Colverson, A.E.	LAC	313058	5	
Combe, G.	F/O		(5)	N.R.V.I.
Comelio, R.B.	LAC	332453	3	20 Squadron
Comerford, H.A.G.	F/O		6	
Compton, G.J.	LAC	332354	3	20 Squadron
Compton, T.W.	LAC	59549	(1)	90 Squadron
Connell, J.	AC2	289142	2,3	

Name	Rank	No.	Clasp	Comments
Conner, S.	AC2	358809	6	
Connery, G.J.	AC1	514193	9	Also 36/37 as LAC
Conrad, S.A.	LAC	359463	6	
Conroy, W.	AC2	508013	6	
Conway, C.P.	AM1	409117	1	
Conway, J.P.	AC1	509293	6,8	
Cook, E.E.	AC2	329348	1,2,3	
Cook, F.C.	AC1	333866	4	
Cook, H.A.J.	AC1	510276	9	
Cook, J.	AM2	300487	1	
Cook, J.H.	AC1	506599	6	
Cook, W.C.	AC2	162323	1	
Cook, W.R.	LAC	341645	6	
Cooke, C.B.	S/L		5	
Cooke, de L.	F/O		6	
Cooke, R.	LAC	46356	6	
Cooksley, C.	LAC	590209	9	Also 36/37 as Cpl
Cookson, S.A.L.	AC2	506329	6	
Cooling, R.M.	Cpl	332179	9	Also 36/37 as Sgt
Coomb, F.J.	AM2	23529	1	
Coombes, A.H.	LAC	362853	6	
Coope, B.R.C.	F/O		5	
Cooper, A.E.	LAC	365694	9	
Cooper, E.E.A.	AC1	511425	9	
Cooper, F.	LAC	226375	1	
Cooper, F.R.J.	Sgt	651	5	
Cooper, G.	Cpl	366071	9	
Cooper, G.H.	AC2	106303	1	
Cooper, H.R.	AC2	345636	4	
Cooper, H.W.	F/Sgt	296365	9	Also 36/37. C/Flt 60 Squadron
Cooper, L.J.	LAC	97942	3	
Cooper, N.J.M.	LAC	365192	6	
Cooper, S.	LAC	561525	9	Also 36/37
Cooper, S.F.	LAC	326987	5	
Cooper, T.	Cpl	341153	5	
Cooper, T.B.	LAC	563070	9	Also 36/37
Copeland, H.G.H.	Sgt	342251	6	
Corbet, L.M.	F/Lt		9	Also 36/37
Corless, C.W.	LAC	514954	9	Also 36/37 as Cpl
Cornelius, H.C.J.	AM1	404200	1	
Cornhill, W.J.	LAC	335901	6	
Corry, A.	AC1	344855	5	
Coryton, W.A.	S/L		4	
Cosh, C.S,	Cpl	352086	6	
Cossum, T.S.	Cpl	350486	9	Also 36/37
Cotsford, E.G.	LAC	513034	9	Also 36/37 as Cpl
Cottle, J.	F/Lt		1	114 Squadron
Couch, T.W.	AC1	344862	4	21/24 in lieu Waz 25
Coulson, W.F.	LAC	506176	6,8	
Coulson, W.G.A.	F/O		9	
Coupe, J.	Cpl	351076	6	
Courtnay-Dunn, A.L.	F/O		2,3	Deceased

Name	Rank	No.	Clasp	Comments
Coveney, G.H.	LAC	365707	9	
Coward, J.G.	LAC	561724	9	
Cowell, A.M.	P/O		6	Forfeiture ?
Cowen, A.V.	SM2	68921	6	
Cowler, W.O.	LAM	47559	1	
Cowley, E.H.S.	AC1	330126	4	
Cowling, W.C.	LAC	561508	9	
Cowling, W.J.	Cpl	340402	6	
Cowton, A.E.	Cpl	342330	4	21/24 in lieu Waz 25
Cox, A.J.	O/O		1,2,3	31 Squadron
Cox, A.O.D.	F/O		9	Deceased
Cox, C.R.	S/L		1,2,3	
Cox, F.	Cpl	357196	6	
Cox, F.W.	LAC	336088	3	20 Squadron
Cox, H.	LAC	561542	9	
Cox, N.C.	LAC	362885	6	
Cox, R.P.	LAC	561489	9	Also 36/37 as Cpl
Coyle, P.J.	AC2	332909	3	20 Sqdn. Not ent. Mahsud 1920
Crabb, S.	AC1	562474	9	Also 36/37 as LAC
Crabb, T.K.	LAC	365402	6	
Crabb, W.N.	LAC	245987	3	
Crabtree, W.	AC2	510616	6,9	
Crackett, H.V.	AC1	370411	6	
Craig, D.H.V.	F/O		6	
Craig, H.J.	Sgt	365622	9	Also 36/37 as F/Sgt
Craig, N.M.	LAC	356520	6	
Craig, T.J.D.	S/L		9	Also 36/37
Craig, W.J.	F/O		9	
Craik, D.	F/Lt		2,3	97 Squadron
Craine, R.D.	Cpl	328280	(5)	N.R.V.I.
Crammond, F.	AC2	342809	3	20 Squadron
Crampton, W.A.	Cpl	353465	6	
Crane, B.	Sgt	6428	5	
Crapper, C.W.	AC2	508049	6	
Craven, A.J.	AC1	156531	1,3	21/24 unclaimed
Crawford, C.W.	Sgt	561511	9	Also 36/37
Crawford, P	AC2	102375	1	
Crawley, J.F.	LAC	363564	6	
Creasey, H.	LAC	358622	6	
Creedon, A.P.	LAC	363367	6	
Cresswell, T.J.A.	Sgt	347225	6	
Crewdon, J.	AC1	515079	9	Also 36/37 as Cpl
Cribb, W.	AC1	340960	4	
Crichton, E.J.	F/O		1	31 Squadron
Crichton, W.M.	AM2	122244	1	
Crisp, P.W.K	LAC	509318	8	
Crisp, R.P.	F/O		1	
Crisp, R.P.A.	Lt		1	
Cro, W.E.	AM2	148737	1	
Crocker, G.E.	LAC	509859	9	
Crocker, L.	F/O		7	
Crockett, W.A.	Sgt	213751	1	

Name	Rank	No.	Clasp	Comments
Croft, E.D.	Cpl	327978	5	
Croft, F.W.	LAC	362287	6	
Croker, E.	AC1	57386	3	20 Squadron
Crompton, V.	Sgt	362854	6	
Cronin, W.D.	LAC	505853	6	
Cronin, W.J.	LAC	365686	9	Also 36/37 as Cpl and 37/39 60 Squadron
Cronkshaw, C.H.	F/Sgt	157724	9	
Crook, E.W.	AC1	362982	6	
Crook, H.C.	AC1	354529	6	
Crook, H.H.	F/Sgt	4345	1	48 Squadron
Crook, R.V.	LAC	362852	6	
Crook, W.E.	Sgt	341879	6	
Croombes, A.H.	Cpl	347149	6,9	Also 36/37 as A/Sgt
Crosbie,	LAM	24042	1	
Cross, H.	Cpl	351266	9	
Cross, H.J.	F/O		6	
Cross, W.C.	AC1	516481	9	
Crossland, J.	AC1	356045	6	
Crouch, A.X.	LAC	563956	9	Also 36/37
Crouch, C.N.	AC1	511412	9	
Crouch, J.N.	LAC	506422	9	
Crouch, R.W	Sgt	313686	3	20 Squadron
Crow, S.	LAC	330695	(5)	N.R.V.I.
Crowdy, F.C	F/Sgt	4042	3	20 Squadron
Crowley, A.P.	LAC	364547	6,8	
Crowley, M.	AM2	164103	1	
Crowther, J.H.	AC1	513488	9	Also 36/37 as Cpl
Cruikshank, G.F.	LAC	358746	6	
Cruise, B.E.	AC1	512146	9	Also 36/37
Crump, L.	AC2	340140	3	20 Squadron
Crunden, P.N.E.	LAC	62441	4	
Cullen, C.F.	LAC	510729	9	Also 36/37 as Cpl
Cullen, L J.	A/Cpl	347045	6	
Culley DSO, S.D.	F/Lt		6,9	
Cumbers, W.R.	Cpl	560327	6,8,9	
Cumming DFC, W.N.	F/Lt		4	21/24 in lieu Waz 25
Cummings, P.H.	F/Lt		3,6	
Cundall, N.	LAC	508841	9	
Cunningham, B.E.	LAC	362833	6	
Cunningham, J. C.	F/O		6	
Curle, W.A.	LAC	563765	9	Also 36/37. 60 Sqdn. Deceased
Curry, R.A.	F/O		1	31 Squadron
Curtis, G.A.	LAC	349427	4	
Curtis, H.	AM1	23264	1	Medal engraved H.V. Curtis AC1
Curtis, J.J.	AC1	508091	6	Not entitled to Mohmand 1933
Cuthbert, E.	O/O		1,2,3	
Cuthell, R.	AM2	49533	1	
Cutler, E.H.	AC1	326637	5	
Dale, A.G.H.	Cpl	51938	4	

Name	Rank	No.	Clasp	Comments
Dale, H.T.	LAC	345061	4	
Dalton, W.	SM	721	4	
Daly, A.P.V.	S/L		4	
Daly, J.A.	Sgt	561100	9	Deceased
Dalzell, F.	LAC	562072	9	Also 36/37 as Cpl
Dand, J,H,	F/O		3	99 Squadron
Daniel, C.B.	Sgt	1242	6	
Daniel, H.McW.	F/Lt		2,3	20 Squadron
Daniel, J.C.	WAC	365712	8	
Daniels, S.	ACI	512106	9	Also 36/37 as Cpl
Dann, F.E.	Cpl	330475	6	
Dann, W.E.	LAC	346936	4	
Darch, H.	AM2	407256	1	
D'Arcy, M.R.	F/O		1	31 Squadron
Dare, P.C.	AM1	16108	1	
Darley CBE AM, C,C,	W/C		6	Also 37/39
Darlington, E,G.	Cpl	366262	9	Also 36/37 as Sgt
Darnbrough, A.H.	Lt		2,3	
Dart, A,	AM2	13618	1	
Darvall, L.	F/O		5	
Dashwood, E.J.	F/O		5	Deceased
Davey, T.A.	AC2	250594	3	97 Squadron
Davidson, A.J.	AC1	342621	(5)	N.R.V.I.
Davidson, A.W.	LAC	350655	6	
Davidson, C.R.	F/Lt		5	
Davidson, E.	Sgt	11990	1	
Davidson, J,	O/O		1,2,3	
Davidson, R.W.	AC2	260316	3	
Davidson, T.J.	F/O		6	
Davidson, W.	LAC	358640	6	
Davidson, W.E.	LAC	359413	6	
Davie, A.E.	AC2	336066	3	20 Squadron
Davies, C.J.	AC1	507478	6	
Davies, C.L.	AM2	148605	1	
Davies, C.T.	F/Sgt	335308	9	Also 36/37
Davies, E.B.C.	F/O		9	
Davies, E.T.	AC1	512957	8,9	
Davies, J.H.	AC2	161570	1	
Davies, P.C.	AC1	508252	6,8	
Davies, S.J.	LAC	365204	9	
Davies, T.A.	AC2	142222	3	
Davis, A.	AC2	516127	9	Also 36/37 as LAC
Davis AFM, D.H.	LAC	562661	9	Also 36/37 as LAC. KIA, MID
Davis, H.A.	LAC	326650	4	
Davis, J.H.	LAC	363587	6	
Davis, L.H.R.	AC2	365710	6	
Davis, M.O.	Sgt	345573	8	
Davis, S.G.	P/O		1	Medal O.C.
Davis, V.H.	Sgt	1058	4	
Davy, P.H.	F/O		2,3	20 Squadron
Dawes, A.	LAC	147304	1	114 Squadron
Dawson, G.H.	AC1	343450	4	21/24 in lieu Waz 25

Name	Rank	No.	Clasp	Comments
Dawson, G.S.	LAC	365400	6	
Dawson, H.	LAC	353279	6	
Dawson, W.	AM2	406509	1	
Day, J.	LAC	79S07	1	31 Squadron
Day, L.B.	AC1	327484	4	21/24 in lieu Waz 25
Deacon, G.	AC2	248890	2,3	
Deacon, G.H.	AC2	257723	1	
Deacon, V.C.	AM2	406580	1	
Dean, A.J.	LAC	149598	4	
Dean, S.G.	F/Sgt	1032	6	
Dear, E.A.	AC2	506974	6	
Dear, R.H.	AC2	245886	3	
Dearlove, C.J.S.	F/Lt		4	
Deaton, W.	LAC	506558	6	
De Boissiere, G.V.	F/O		1	
De Fontenay, P.A.	F/Lt		4	Cashiered
Delamain, E.C.	F/O		1,2,3,6	
Delany, C.J.A.	F/O		5	
De Moyse-Bucknall, S.P.B.	P/O		1,2,3	
Denham, E.T.	AC2	198209	(2),(3)	N.R.V.I.
Dennett, J.R.	Sgt	340921	4	
Dennett, W.	A/Cpl	514915	9	Also 36/37 as LAC
Dennis, J.J.	AC1	341762	4	
Dent, A.A.G.	LAC	560596	9	
Dent, C.H.	AC2	329264	3	20 Squadron
Dent, J.	AC1	359307	6	
Dent, P.W.E.	AC1	366078	6,8,9	
Dent, R.C.	Cpl	359091	8	
Dert, C.M.	AC2	329264	1	
Deslandes, H.J.	AC2	245293	3	NWF 1919 & Mahsud N.R.V I.
De St. Leger, R.J.M.	F/Lt		4	
Desoer, N.L.	F/Lt		6	
De Souza, S.A.J.	LAC	238327	3	99 Squadron. Not ent. 21/24
D'Este, H.A.M.	Capt		6	
De St. Croix, P.E.	LAC	364074	6	
Devan, A.	LAC	365436	6	
Devaney, A.W.	AC2	55266	1	
Devas, W.G.	F/O		9	
Deveria, F.A.	AM3	103158	1	
Devon, W.	LAC	509067	9	
Devonish, W.G.	LAC	358696	6	
Devonshire, F.V.	F/O		1	31 Squadron
Dew, W.	AC2	329701	3	20 Squadron
Dewar, R.	LAC	562506	9	
Dewson, T.A.	LAC	17189	1,3	20 Squadron
Dick, W.A.	AC1	507044	6	
Dicken, A.	LAC	365221	9	
Dickenson, S.	LAC	345720	4	
Dickie, H.	LAC	344620	5	
Dickson, B,W,C,	AC1	363576	6	
Dickson, G.C.	AC2	191757	1,2,3	
Diggins, R.J.	AC2	3341R1	3	

Name	Rank	No.	Clasp	Comments
Dillamore, A.	ACl	157727	4	
Dillon, R.W.	LAC	505776	6	
Dimbylow, H.W.	LAC	326471	4	
Dinham, W.E.	Sgt	332236	6	Deceased
Dinnage, H.A.	F/0		1	114 Squadron
Dipple, W.E.	F/Lt		6	
Dix, S.B.	LAC	359355	6	
Dixon, A.R.	Sgt	38947	1	
Dixon, F.E.R.	P/O		5	
Dixon, J.W.	LAC	363913	6	
Dobie, T.	Cpl	560750	9	
Dobson, G.	AM2	21189	1	
Dobson, J.	LAC	561686	9	
Dobson, J.	LAC	513029	9	
Dod, H.H.	AC1	326201	4	
Dodd, G.F.	Cpl	346656	9	Also 36/37 as Sgt
Dodd, R.	ACl	513854	9	
Dodds, F.G.	AM2	409134	1	
Dodds, G.W.	F/Sgt	239144	9	Also 36/37
Dodds, T.D.	LAC	342689	4	
Dodkins, L.C.	F/0		1	Deceased
Dodridge, C.B.	Cpl	560095	9	Also 36/37 as Sgt
Dodson, J.A.	AC2	158238	3	
Doig, R.	AC1	349067	(5)	N.R.V.I.
Dole, W.F.	LAC	363382	6	
Dollery, C.	F/0		4	21/24 in lieu Waz 25
Dolphin, S.W.	AC2	240766	3	20 Squadron
Dolphin, W.H.	F/Lt		1,2,3	
Don, D.S.	F/Lt		3,4	1st Wing att. 20 Squadron
Donaldson, A.	LAC	359694	6	
Donaldson, G.	LAC	358520	6	
Donaldson, G.F.K.	F/O		8	
Donavan, C.W.	Cpl	361943	6	
Donbavand, N.R.	LAC	363075	6	
Donnan, W.G.	LAC	250618	2,3	
Donovan, J.	SM1	27407	1	
Doody, V.A.	LAC	244904	3	
Dormor, R.B.	F/O		1	Medal O.C.
Dorrell, B.A.	LAC	509983	8,9	
Double, A.E.	AC1	505798	6	
Douche, F.A.B.	AC1	370497	6	
Dougall, N.S.	O/O		2,3	
Dougan, D,	AC2	116147	3	
Dougherty, C.L.	AC2	342629	3	20 Squadron
Doughty, P.A.	AC1	354503	6	
Douglas, J.S.	F/O		6,8	
Douglas EGM, R. E.	LAC	362370	6	MID. 60 Squadron
Douglas, R.P.	AC1	326868	5	
Dower, J.H.	LAC	359292	6	
Dowle, W.H.	LAC	107500	2,3	20 Squadron
Dowling, W.H.	Lt		1	114 Squadron
Downer, W.J.	LAC	365725	6	

Name	Rank	No.	Clasp	Comments
Downes, T.H.	F/Lt		9	
Doyle, T.J.	AC2	342857	3,4	20 Squadron
Drake, A.V.	AM3	409047	1	
Drake, H.L.	F/O		3	Medal engraved "Lt H.L. Drake 28th Punjabis"
Drake, W.	AM2	21128	1	
Drew, G.A.F.	AC2	269603	3	99 Squadron
Drew, H.E.	LAC	365226	6	
Drew, J.W.N.	LAC	370349	6	
Drew, V.B.W.	AC1	363392	6	
Drewitt, S.G.	Sgt	335380	6	
Dring, A.E.	LAC	370037	6	
Driver, A.	LAC	327464	5	
Drudge, G.F.	F/O		3	
Druett, J.	AM2	409048	1	
Drury, K.G.	LAC	364614	8,9	
Drury, R.C.	F/Sgt	329179	9	Also 36/37 as W/O
Drummond-Hay, P.	AC2	247028	1	
Dudley, W.A.L.	Sgt	243182	3	
Duff, N.H.	AC1	511487	9	Also 36/37 as Cpl. Deceased
Duffan, F.	AC1	343012	5	
Duffey, H.W.	F/O		6	
Duffy, S.E.	LAC	561102	9	Also 36/37
Duggan, L.B.	F/O		1,2,3	Medal O.C. 21 and 99 Squadrons
Dukes, G.E.	LAC	352937	6	
Dummett, F.C.	AC2	132662	1,2,3	
Dunbar, J.C.	F/O		2,3	5 Squadron
Duncan, A.	LAC	349054	3	
Duncan, D.C.	F/O		1,2,3	99 Squadron. Deceased
Duncan, J.	AC2	341326	3	20 Squadron
Duncan, O.T.	LAC	342314	5	
Dunford, D.D.	LAC	561106	9	Also 36/37
Dunford, E.J.	AC2	9466	5	
Dunkinson, K.J.	Cpl	340464	6	
Dunkley, A.G.H.	LAC	363585	6	
Dunn, W.H.	F/Lt		3	
Dunnett, D.	AC2	340779	4	21/24 in lieu Waz 25
Dunson, T.L.	LAC	340592	4	
Dunstan, C.	AC2	344192	4	
Dunstan, F.	AC1	343241	3,5	20 Squadron
Dunworth, M.J.W.	AC1	364071	6,8	
Durbridge, G.W.K.	LAC	245642	4	
Durham, J.S.	Cpl	334504	4	
Dust, R.A.	LAC	364612	9	
Dutton, S.F.E.	Sgt	336059	3	31 Squadron att. 20 Squadron
Dutton, P.H.	F/Lt		9	Also 36/37 & 37/39. Deceased
Dutton, R.H.	Cpl	346659	4	
Dwyer, J.A.	LAC	359428	6	
Dye, R.	F/Sgt	3405	3	
Dyer, N.E.	LAC	355993	6	
Dymond, A.E.	AC1	370676	6	
Dyne, A.V.	LAC	357453	6	

Name	Rank	No.	Clasp	Comments
Eades, L.A.	LAC	180029	(5)	Also 36/37 as Sgt
Eady, T.W.G.	F/O		6	
Eagar, R.A.	Cpl	343194	9	
Eager, F.J.	LAC	361960	6	
Early, O.S.	LAC	408845	(1)	Attached 1/7 Brigade RFA
East, E.S.	LAC	363870	6	Deceased
Eastaway, E.A.	Cpl	356081	6	Formerly V.A. Hogan
Eastley, C.M.	F/Lt		1	31 Squadron
Easton, E.C.	Cpl	358711	6	
Easton, G.E.	Cpl	342252	6	
Easton, J.W.	Cpl	348589	6	
Eaton, W.A.	LAC	511089	9	
Eayrs, D.J.	F/O		6	
Eccles, A.T.	LAC	348728	5	
Eckett, W.J.	Cpl	347772	9	
Edeveain, F.G.	AC2	508720	6	
Edgar, I.G.G.	F/O		2,3	
Edge, F.	AC1	62160	1	
Edge, J.A.	LAC	507687	9	
Edge, T.W.	LAC	370105	6	
Edington, A.	LAC	366092	9	Also 36/37
Edmonds, A.	AM2	406395	1	
Edmondson, W.S.	AM	63107	1	
Edmunds, D.W.	LAC	513935	9	Also 36/37 and 37/39
Edwards, C.J.	Cpl	361823	6	
Edwards, E.C.		410431	1	Now 5430817 Sgt DCLI
Edwards, E.D.	F/O		1	20 squadron
Edwards, F.J.M.	AC1	515539	9	Also 36/37
Edwards, J.E.	Cpl	346379	5	
Edwards, O.G.	LAC	348589	4	
Edwards, W.	AC2	510647	6,9	
Edwards, W.A.F.	LAC	515150	9	Also 36/37. 60 Squadron
Eggleston, C.C.	W/O	53302	9	Also 36/37
Elder, J.M.	LAC	361876	6	
Eldridge, H.J.	Cpl	349337	6	A/Sgt
Eldridge, S.G.	LAC	513262	9	Also 36/37
Eldridge, W.G.	LAC	315022	5	Deceased
Eley, J.S.	SM2	845	3	
Elias, A.W.	F/Lt		9	Deceased
Ellaby, T.H.	LAC	361957	6	
Ellcock, T.W.	AC2	239399	1,2,3	Medal O.C. 20 Squadron
Ellen, C.N.	O/O		2,3,8	31 Squadron
Ellett, C.H.	AC1	511417	8	
Elliot, G.A.	F/O		3	20 Squadron
Elliott, A.	F/Sgt	190197	9	Also 36/37
Elliott, C.A.	F/O		1,3,4	52 Wing
Elliott, G.A.R.	Cpl	249264	9	Also 36/37 as Sgt
Elliott, J.A.	F/O		3	97 Squadron
Elliott, J.W.	LAC	512510	9	
Elliott, T.H.	LAC	282219	6	
EllLott, W.M.C.	LAC	362326	6	

Name	Rank	No.	Clasp	Comments
Ellis, C.	LAC	370357	6	
Ellis, E.T.H.	F/0		3	
Ellis, G.H.	Cpl	10764	1	
Ellis, H.C.	AM2	404270	1	
Ellis, J.H.	LAC	363071	6	
Ellis, R.W.	LAC	363389	6	
Ellis S,G.	LAC	327388	5	
Ellis, W.	AC1	2416S9	1	
Ellison, E.E.	F/0		6	
Ellison, R.C.W.	F/O		6	
Ellwood, A.B.	F/Lt		4	
Elms, M.B.	AM2	241518	1	
Elsom, L.	AC2	284237	1,3	
Elvy, A.	LAC	328002	4	21/24 in lieu Waz 25
Elwell, G.H.	AC2	342359	3	20 Squadron
Embry AFC, B.E.	F/Lt		( )	Also 36/37 and 37/39
Emburey, F.T.	LAC	356217	6	
Emeny, G.W.	Cpl	364636	6	
Emery, H.V.	AC1	247230	(1),3	20 Squadron
Emmerson, A.B.	AC1	354156	6	
Emmerson, W.L.	LAC	561534	9	
Endacott, C.	AC2	516243	9	Also 36/37 as AC1 (A/Cpl)
England, G.J.	AC2	329996	3	20 Squadron
England, J.H.	LAC	359208	6	
English, J.J.	AC1	506197	6,8	
Etheredge, C.G.	Cpl	353489	9	
Ettle, J.W.	LAC	336849	4	
Evans, A.E.	F/Lt		9	Also 36/37
Evans, D.T.	AC2	334296	3	Not entitled to 21/24
Evans, F.W.	LAC	507312	9	
Evans, G.T.	LAC	345349	5	
Evans, H.	AC2	163301	3	20 Squadron
Evans, L.S.	LAC	506246	6	
Evans, R.	LAC	509686	9	
Evans, R.G.	AC1	247411	3	20 Squadron
Evans, W.J.	LAC	370175	6	
Evans, W.M.	AC1	364643	6,8	
Eveleigh, S.A.	Cpl	351062	8,9	A/Sgt
Evenett, R.A.	Cpl	505971	9	
Everett, E.N.V.	F/O		6	
Everett, W.F.	LAC	361706	6	
Everidge, J.	S/L		1	
Everitt, A.V.	LAC	358726	6	
Everitt, E.	AC1	333036	3	99 Squadron
Everitt, W.C.G.	LAC	513993	9	Also 36/37. 60 Squadron
Eversfield, H.M.T.	F/0		9	60 Squadron
Ewens, F.G.H.	F/St		9	Also 36/37 as S/L
Ewing, W.G.H.	F/O		6	
Eyre, J.A.	Sgt	364653	9	
Fagg, H.J.	LAC	343548	(5)	N.R.V.I.

Name	Rank	No.	Clasp	Comments
Fairbairn, W.R.	F/O		2	
Fairbank, G.	Cpl	345608	6	
Fairbrother, F.A.J.	LAC	562113	9	Also 36/37
Fairman, F.W.	AC1	402569	1	
Fairweather, J.	AM2	148786	1	
Fairweather, J.MacG.	S/L		9	
Falconer DFM, R.A.R.	Sgt	363944	6	
Fallows, H.	Cpl	356803	9	
Fancy, A.S.	LAC	366275	9	Also 36/37
Fancy, C.H.W.	LAC	364647	9	
Fancy, G.E.	AC2	345965	4	
Farebrother, F.T.	AC1	506928	8	
Farlowe, G.E.	Cpl	361962	6	
Farmer, A.	Sgt	363945	6	
Farmer, F.J.E.	LAC	363399	6	
Farmer, F.W.	AM2	47583	1	
Farmer, H.	AC1	139211	2,3	20 Squadron
Farmer, J.C.P.	AC1	561535	9	
Farnsworth, T. H.	Sgt	365630	7	
Farr, J.W.			1	20 Squadron
Farran, S.	A/Sgt	93924	6	
Farrell, L.E.	Sgt	364019	8,9	
Farrer, A.A.	Cpl	340363	9	
Farrington, W.B.	W/C		9	
Farrow, C.R.	Cpl	344703	1,6	Medal and clasp Afg 1919 issued for Army Service
Farrow, G.S.	LAC	562672	9	Also 36/37
Faulkner, E.M.	LAC	505932	6	
Faver, V.	AC2	350091	4	
Fawcett, J.	LAC	63613	1	
Fawcett, W.	LAC	560113	9	
Fay, A.J.	AC1	511456	9	
Fearn, W.H.	LAC	328556	4	
Feather, C.	F/Lt		6	
Featherstone,	Lt		1	
Featherstone, M.	AC2	248402	1	
Felix, A.L.	Cpl	343048	4	
Fells, L.T.	LAC	350302	5	Waz 25 in lieu 21/24
Fenton, E.N.	F/O		2,3	
Fenton, H.A.	F/O		6	
Fenton, S.W.	AC1	189053	2,3	1st Wing
Fenwick,	AM1	80629	1	
Fereday, S.H.	Sgt	6409	1,2,3	
Ferguson, J.	Cpl	329440	9	Also 36/37
Ferrier, F.G.	F/O		6	
Ferris, S.	AC2	290520	1,3	Medal O.C.
Ffoulkes-Jones, E.J.	F/Lt		6	
Fiddament, A.L.	F/O		2,3	20 Squadron
Fiddick, C.A.	Cpl	363134	9	
Field, E.W.	AC2	507167	6	
Field, J.W.A.	LAC	337051	3	52 Wing
Field, R.A.F.	AC2	298601	3	

Name	Rank	No.	Clasp	Comments
Field, W.	Sgt	13928	1	
Fielden, N.	F/O		1,2	
Fielder, R.W.	LAC	354498	6	
Fielding, A.	AC1	513346	9	Also 36/37
Figg, A.	W/O	19797	9	
File, F.G.	AM1	16733	1	
Finch, F.	AM2	410303	1	
Findlay, M.H.	F/O		2,3	
Fippard, N.A.	LAC	365408	6	
Firth, E.	LAC	362508	6	
Fisher, A.	Cpl	327793	4,8,9	Also 36/37 as W/O
Fisher, A.	LAC	511668	9	
Fiske, A.W.	AC1	342698	3,(5)	20 Squadron. N.R.V.I.(Waz 25)
Fitchew, E.E.	LAC	562095	9	
Fitt, P.A.	Cpl	353668	6	
Fitton, E.C.	Sgt	2954	1	
Fitzgerald, J M.	AC1	158997	2,3	
Fitzgerald, T.G.	LAC	508785	9	Also 36/37
Fitzgerald-Lee, G.	LAC	363516	6	
Flaxman, C.A.	LAC	361966	6	
Fleet, E.J.	LAC	365741	6	
Fletcher, A.E.	AC2		1	5 Squadron
Fletcher, F.E.	Cpl	143668	6	
Fletcher, G.	Cpl	348190	9	
Fletcher, H.	AC1	350429	4	
Fletcher, J.R.	AC2	149463	1	
Fletcher, R.J.	LAC	64011	3	20 Squadron
Flint, G.F.J.	LAC	335403	6	
Flint, M.C.W.C.	F/O		5	
Flitter, W.E.	AC1	505959	6,8	
Flood, W.	LAC	363586	6	
Florence, F.	Sgt	303543	8,9	
Fluellen, F.L.	Cpl	24203	5	Also took part in 21/24 Opns
Foley, J.	AM2	407602	1	
Foord, B.A.	Lt		1	
Foot, C.G.	LAC	355073	6	
Forbes, A.K.	F/Sgt	3719	4	
Forbes, W.J.	LAC	365851	9	
Forbes-Mitchell, W.H.	F/O		9	
Ford, A.H.	Sgt	84737	6	
Ford, H.	F/O		4	
Ford, S.J.	LAC	244136	4	21/24 in lieu Waz 25
Forde, J.F.	Cpl	352062	6	
Forman, R.	LAC	17523	1	
Forrester, J.S.	Cpl	344875	6	
Forshaw, J.	Cpl	356712	6	
Forster, A.	Sgt	43706	1	
Forster, J.H.	AC1	359595	6	
Forster, J.P.	LAC	336542	3	52 Wing
Forsyth, E.	LAC	370221	6	Also 36/37 as Sgt
Forsythe, C.McM.	LAC	512258	9	A/Cpl
Forsythe, J.M.	AM1	17631	1	

Name	Rank	No.	Clasp	Comments
Forth, R.T.	Cpl	341894	5	Also took part in 21/24 opns
Fortune, R.B.	P/O		2,3	
Forward, I .	Sgt	361902	6	
Foster, C.E.W.	F/Lt		1,2,3	Medal O.C.
Foster DFC DSM, F.W.	F/O		6	
Foster, G.P.	AM2	62421	1	
Foster, H.E.	2/Lt		1,2,3	
Foster, H.J.	Cpl	347272	6	
Foster , R. M.	F/O		2,3,4	20 squadron
Foster, W.	Cpl	338023	5	
Foster, W.E.	LAC	355985	6	
Fowle, W.F.	AC1	511823	5	Also 36/37
Fowler, S.A.	AS1	136412	1	
Fox, A.	A/Cpl	26505	1 , 3	
Fox, B.A.	LAC	334030	4	
Fox, C.H.	AM2	148728	1	
Fox, F.	AC1	356435	6	
Fox, G.P.	AC2	308521	3	
Fox, J.B.	F/Lt		3	Deceased
Fox, J.C.	SM	6482	1	
Fox, J.M.	AM2	410347	1	
Fox, R.C.	AC1	340915	4	
Fox, R.T.E.	AC1	82798	4	
Fox, S.C.	AC2	158712	3	20 Squadron
Foyle, E.J.	Cpl	342593	5	
Francis, A.S.	AC1	505778	6	
Francis, C.W.	A/Sgt	349822	9	Also 36/37 as F/Sgt
Frankland, S.V.	AC1	506647	6	
Franks, J. G .	F/Lt		6	
Frankum, W.	AC1	342325	4	21/24 in lieu Waz 25
Fraser, S.E.	Cpl	328191	9	
Fraser, S.F.	Cpl	877	1,2,3	28 Squadron
Frazier, B.	LAC	333786	4	
Freathy, E.J.	LAC	361965	6	
Freeman, J.H.	AC1	358019	6	
Freeman, O.D.	F/O		4	Medal lost. Dup issued 27.2.28
Freeman, W.R.	Cpl	349804	4	
French, E.H.	Sgt	331753	9	
French, H.	Sgt	329908	9	
Fressanges, F.J.	F/Lt		6	
Frewin, E.F.	AC2	509161	6	
Frewin, T.H.	LAC	505216	6	
Frisby, W.J.	LAC	60120	1	
Frith, G.	AC2	507297	6	
Frost, J.W.	LAC	355235	6	
Frost, R,	Cpl	560119	9	
Frost, W.J.	F/Sgt	244430	9	Also 36/37
Fry, E.	F/Sgt	51126	1	
Fugler, W.R.J.	Cpl	4358	6	
Fulcher, W.	LAC	354623	6	
Fulford, A.A.J.	LAC	364082	8,9	Also 36/37
Fullard, P.F,	S/L		9	

Name	Rank	No.	Clasp	Comments
Fuller-Good, J.L.F.	F/Lt		6	
Fullick, E.C.	AC1	359534	6	
Fulton, R.	AC1	359227	6	
Fulwell, C.	L/Cpl	3308239	9	H.L.I. (Army)
Furnell, W.H.	AC1	326086	4	21/24 in lieu Waz 25
Fursman, A.	AM2	111098	1	
Fyfe, D.	AC1	342213	4	21/24 in lieu Waz 25
Fyfe, J.B.	F/O		6	
Gadd, E.	AC1	330345	5,9	Also 36/37
Galaud, H.J.A.	LAC	363606	6	
Gale, A.	AC1	509577	9	
Gale, T.G.L.	F/O		9	
Gallagher, A. F.	AC1	343619	4	
Gallie, J.	Sgt	122404	6	
Galloway, R.G.	LAC	513834	9	Also 36/37
Gamblen, E.R.	LAC	364070	6	
Game, M.	LAC	510599	9	Also 36/37
Gammack, D.J.	F/Sgt	341552	6	
Gandy, H.	Cpl	562687	9	
Ganley, L.C.	LAC	362342	6	
Gape, H.I.	AC2	337960	3	20 Squadron
Gardiner, E.J.	LAC	158190	4	21/24 in lieu Waz 25
Gardiner, T.	LAC	250696	1	20 Squadron
Gardiner, W.A.	Cpl	344486	9	
Gardner, C.	F/O		5	
Gardner, D.W.H.	F/O		9	Also 36/37
Gardner, L.J.	LAC	365739	9	
Garland, A.H.	F/O		9	
Garlick, J.E.	LAC	561732	9	
Garner, A.	Cpl	7896	1	
Garner, J H.	AC2	113700	1	
Garner, R.H.	F/Sgt	254122	1,2,3	
Garnett, A.E.	LAC	363605	6	
Garnett, E.	LAC	344850	5	
Garraway, F.F.	F/O		5	
Garrett, E.L.W.	Cpl	326535	9	
Garrett, H.	AC1	158716	5	
Garrett, N.A.C.	LAC	365426	6	
Gartery, D.M.	LAC	562388	9	Also 36/37
Garton, G.E.	AC2	363875	6	
Gasson, A.G.	F/Sgt	332746	6	
Gaston, V.	Cpl	332948	3	20 Squadron
Gates, H.E.	LAC	362360	6	
Gaussen, F.D.D.	P/O		1	
Gay, G.W.	F/O		4,6	
Gaynor, W.F.	LAC	363100	6	
Geal, A.J.	LAC	506484	6	
Geal, F.T.	LAC	136895	1,2,3	
Gentry, F.C.	Cpl	335512	6	
George, A.	LAC	354992	6	

Name	Rank	No.	Clasp	Comments
George, E.F.	LAC	361976	6	
George, R.A.	F/O		3,4	97 Squadron
George, S.P.	F/O		(5)	N.R.V.I.
Gibbon, J.H.D.H.	LAC	365859	6	
Gibbons, G.J.	Cpl	351715	9	
Gibson, A.O.	Cpl	335418	9	
Gibson, D.B.	AM2	75221	1	
Gibson, L.J.	AC1	365848	5	Not entitled to Mohmand 1933
Giddens, N.P.B.	P/O		1	
Gilbert, A.J.	Sgt	341318	4	
Gilbert, L.	Sgt	8974	6	
Gilbert, T.	F/O		1	114 Squadron
Gilbertson, G.C.	LAC	353057	4	
Gilhome, W.H.	LAC	357187	6	
Gilks, W.H.	LAC	563708	9	Also 36/37
Gill, G.L.	Cpl	364066	9	
Gill, O.A.	LAC	561550	9	Also 36/37
Gill, R.H.	LAC	363128	6	
Gilling, R.E.	Cpl	48688	1,3	
Gillman, J.R.	P/O		9	
Gilloly, B.A.	AM3	409143	1	
Gilmartin, W.R.	LAC	356426	5	A/Sgt
Gilmour, R.P.	LAC	364063	6,8	
Gilmour, W.J.	LAC	511601	9	Also 36/37
Girling, A.N.	LAC	363104	6	
Girling, H.V.	AC2	365223	6	
Glanfield, B.	AC1	348527	6	
Glass, S.	AC1	341944	4	Also took part in Waz 25 Opns
Glasscock, E.A.	LAC	359395	6	
Gleave, G.	AM1	9558	1	
Glenfield, B.	LAC	342257	5	
Glenny, A.W.F.	F/Lt		2,3,4	20 Squad. Also 37/39. Deceased
Glover, C.H.	F/O		6	
Glyde, J.B.V.	F/O		4	
Goacher, A.M.	Sgt	4414	1	
Goat, G.W.	F/Sgt	157193	6	
Goater, A.W.	LAC	342235	(5)	N.R.V.I.
Godby, R.G.R.	F/O		4	
Goddard, F.A.		13645	I	
Godfree, H.V.	LAC	363396	6	Also 36/37
Godfrey, D.	LAC	364029	6	
Godfrey, J.	Cpl	363970	6	
Godfrey, J.	LAC	112342	I	
Godfrey, J.M.	F/O		1,(2),(3)	
Godfrey, L.E.G.	LAC	335639	6	
Godfrey, R.H.H.	AC1	357387	6	
Godsell, R.A.	Cpl	363932	8	
Goffe, D.W.	LAC	365833	6	
Goldfinch, G.	AC2	221142	1	
Goldfinch, S.G.E.	AC1	221154	3	
Golding, C.A.	AC1	509633	8	
Golding, F.J.	AC2	248038	5	

Name	Rank	No.	Clasp	Comments
Goldsmith, F.L.	LAC	333097	3	31 Squadron att. 20 Squadron
Goldsmith, N.D.	AC2	86643	3	20 Squadron
Goldstone, G.J.	Cpl	342248	4	Also took part in Waz 25 Opns Also 36/37
Gomm, W.J.T.	LAC	248037	4	Also took part in Waz 25 Opns
Gooch, A.E.	LAC	506685	6	
Goodall, R.	LAC	357972	6	
Goode, E.J.H.	LAC	357559	6	
Goode, H.	LAC	509624	6	
Gooding, C.	AM2	148739	1	
Goodman, A.	Cpl	354928	6	
Goodridge, G.H.	Cpl	560148	9	
Goodridge, W.C.R.	AC1	364642	6	
Goodsell, W.E.	LAC	562705	9	Also 36/37
Goodwin, A.S.	F/Lt		(3),(4)	N.R.V.I.
Goodwin, H.F.	LAC	342787	5	
Goodwin, W.J.	Cpl	12395	1	T/Sgt
Gordon, G.	AM1	148723	1	
Gordon, J.G.	LAC	365838	9	
Gordon, R.B.	0 / 0		3	Deceased
Gordon, R.C.L.W.	Sgt	335610	6	
Gordon, S.	Cpl	156537	3 , 4	
Gordon, W.	LAC	341876	5	
Gordon-Dean AFC, H.	W/C		6	
Gore, C.W.	F/Lt		9	Also 36/37
Gorring, A.	AC1	370776	6	
Gosling, F.A.	LAC	590036	8	
Gosling, J.P.	AC1	363958	6	
Gosnell, R.J.	F/O		9	
Goss, F.	LAC	364798	9	
Goss, L.B.	Sgt	80749	3	52 Wing
Goudey, J.E.	P/O		1 , 2 , 3	
Goudge, B.N.	Lt		1	
Gough, T.H.	AC1	512487	9	Also 36/37
Gough, W.J.	LAC	333613	3	20 Squadron
Gould, S.H.	LAC	562719	9	
Gould, S R.	AC1	330527	5	
Gouldby, A.D.	AC1	506189	6	
Gondy, W.E.	Cpl	364065	6	Deceased
Gowing, H.J.	Sgt	363612	6	
Gowland, D.G.W.	LAC	505862	6	
Grace, C.F.H.	F/O		4	
Graham, C.T.	Sgt	335739	9	Also 36/37
Graham, J.B.	S/L		3	
Graham, J.H.	LAC	506906	9	
Graham, J.S.	Sgt	562115	9	Also 36/37
Graham, R.	F/Lt		3	
Graham, S.	F/Lt		5	
Graham, W.	AC1	513043	9	Also 36/37
Graham,W.J.C	LAC	365228	9	
Graham, W.M,	F/O		9	Also 36/37. 27 Squad. KIA 1940
Grant, A.H.G.	LAC	365836	9	Also 36/37 as Cpl

Name	Rank	No.	Clasp	Comments
Grant, F.	AC2	298050	3	
Grant, R.	AM2	18471	1	
Grant, W.V.N.	Lt		1	48 Squadron
Grantham, F.	AC1	515209	9	Also 36/37
Graves, J.	AC2	341878	5	
Gray, A.A.	LAC	363855	6	
Gray, E.E.	AC2	101528	3	
Gray, E.N.H.	S/L		3	
Gray, F.	LAC	341564	4	21/24 in lieu Waz 25
Gray, F.	Sgt	351960	9	
Gray, G.	LAC	342846	5	
Gray, H.E.P.	Cpl	363407	6	
Gray, J.	LAC	560786	9	
Gray, R.H.	LAC	563555	9	A1so 36/37
Gray, S.H.C.	F/O		6	
Gray, W.McL.	Cpl	177364	6	
Grayson, F.S.	LAC	370076	6	
Green, C.T.H.	LAC	364026	6	
Green, A.N.	AM2	49033	1	
Green, E.S.	AC1	509911	9	
Green, F.E.	LAC	560156	9	
Green, H.	LAC	343728	4	
Green, H.S.	F/O		1	
Green, J.A.	LAC	365281	9	Also 36/37 as Cpl
Green, K.A.	LAC	364812	6	
Green, T.H.	AM2	407609	1	
Green, W.H.	LAC	352150	6	
Green,W.J.	Cpl	335896	6	
Creener, J.	SM1	737	5	
Greenfield, A.E.	AC1	505832	6	
Greenhalgh, J.R.	LAC	510537	9	
Greenhall, A.	AM1	35579	1	
Greenhill, H.C.	Sgt	352215	8	
Greenland, E.T.	LAC	363403	6	
Greenoff, F.W.	Sgt	350311	6	
Greenslade, R. S.	F/O		1, 2 ,3	31 Squadron. Medal O.C.
Greenslade, T.D.	AC1	363067	6	Also GSM (S,Desert Iraq)
Greenslade, W.B.	Sgt	329164	6	
Greentree, A.E.	LAC	335303	6	
Greenwood, E.J.	Sgt	560760	9	Also 36/37
Greenwood, E.	Major		3	O/O
Greenwood, S.	Sgt	18473	3	
Greet, F.I.	LAC	362935	6	
Gregory, G.W.	AC1	48845	1	
Gregory, V.G.D.	LAC	560173	9	Also 36/37 as Cpl
Grellis, H.E.	LAC	362113	6	
Grenfell, E.O.	F/L		3,4	Medal O.C.
Grierson, J.	P/O		6	Forfeiture ?
Grieve, C.D.	AC1	511582	8,9	Also 36/37 as Cpl
Grieve, R.	AC1	512036	9	A/Cpl
Griffin, A.J.	AC2	337367	3	52 Wing
Griffin, E.H.	AC1	370020	6	

Name	Rank	No.	Clasp	Comments
Griffin, L.T.M.	Lt		1,3	52 Wing
Griffin, N.J.	LAC	563663	9	Also 36/37 as Cpl
Griffiths, A.	LAC	327823	4	
Griffiths, A.H.	LAC	245777	5	
Griffiths, E.	LAC	561866	9	
Griffiths, J.H.	Cpl	346855	9	Also 36/37 as Sgt
Griffiths, J.V.	Sgt	14686	1	
Griffiths, W.N.C.	Sgt	365730	6	
Grigg, F.	Cpl	345676	5	
Grigson DSO DFC, J.W.B.	W/C		9	Also 36/37. KIA 3.7.42
Grogan, G.J.	LAC	560970	9	
Groom, F.E.	F/Sgt	327338	9	
Groome, M.L.	AC1	343999	4	Also took part in Waz 25 Opns
Grosclaude, J. H.	Cpl	512671	9	Also 36/37 as Cpl
Grout, D.L.	Sgt	363107	6	
Groves, S.	AC2	505891	7	
Groves, T.E.	LAC	326138	5	
Grundy, W.H.	LAC	358420	6	
Guille, L.	AC2	238714	1	
Guillermin, G.H.	AM1	404285	(1)	Attached 31 Squadron
Guiver, H.	LAC	358487	6	
Gulliver, C.L.	LAC	364062	6	
Gunn, A.E.	Cpl	507221	9	
Gunn, J.	Cpl	340149	3	
Guy, J.C.	LAC	359278	6	
Guy, J.W.	AC1	513711	9	Also 36/37 as Cpl
Guyton, P.G.	Sgt	245896	3	
Habberfield, F. G.	LAC	562996	9	Also 36/37
Hackett, C.E.J.	LAC	364826	6	
Hadaway, E.H.P.	LAC	561944	9	
Hackney, J.R.	LAC	561730	9	Also 36/37
Hadden, H.J.	AC1	327349	4	
Hadley, V.J.S.	AC2	505885	6	
Hadley, J.W.	Cpl	365903	9	Also 36/37
Haigh, C.R.	Sgt	99730	1 , 3	
Hailes, J.G.	LAC	362971	6	
Haimes, H.E.	AM2	5376	1	
Haines, G.J.W.	AC1	506310	6	
Hakes, T.	AC1	328046	(5)	N.R.V.I.
Hale, A.W.J.	Sgt	361977	6	
Hale, T.J.	AC1	340631	4	
Halford, W.	F/O		3	5 Squadron
Hall, D.E.	P/O		1	Medal O.C.
Hall, F.	LAC	355881	6	
Hall, F.F.W.	F/Lt		6	
Hall, H.E.	LAC	341752	4	Also took part in Waz 25 Opns
Hall, J.	AC1	370474	6	
Hall, J.	LAC	510905	9	Also 36/37
Hall, R.E.	F/Lt		6	
Hall, R.W.M.	F/Lt		5	

Name	Rank	No.	Clasp	Comments
Hall, T.R.	LAC	560319	9	
Hall, W,F,	LAC	364821	6	
Hall, W.H.	AC1	54902	(1)	5 Squadron
Hallam, C.F.	AC2	343174	3	20 Squadron
Hallett, W.T.	W/O	111396	9	Also 36/37
Hallett, L.A.	LAC	560294	9	Also 36/37
Halley, H.V.	AC1	261556	3	20 Squadron
Halley, R.	F/Lt		1	31 Squadron
Halliday, C.C.	LAC	506687	8	
Halstead, A.	F/Sgt	314223	5	
Halwood, H.	F/Sgt	4337	3	31 Squadron att. 20 Squadron
Hambleton, H.F.	AC2	505896	6	
Hamblin, R.K.	F/Lt		9	Also 36/37 as S/L
Hambly, F.McG.	LAC	365370	9	Also 36/37
Hamersley, H.A.	F/Lt		4	
Hamilton, L.S.	F/Lt		5	
Hamilton, R.	O/O		3	
Hamley, P.H.	F/O		9	
Hammock, G.H.	Cpl	242990	4	
Hammond, F.G.	F/Sgt	6606	1,2,3	
Hampshire, A.S.T.	LAC	359728	6	
Hampshire, W.H.C,	Sgt	342755	5	
Hampton, A.H.	Cpl	353477	6	
Hancock, C.R.	F/Lt		6	
Hancock, P.V.	AC2	333735	3	20 Squadron
Hancock, R.C.	F/O		6	
Hancock, R.W.	F/Sgt	247680	6	
Hand, H.G.	LAC	364831	9	
Handel, E.	LAC	357957	6	
Handley, A.S.	AC1	328080	4	
Hanglin, A.	W/O	4233	9	
Hannigan, W.H.	LAC	561728	9	Also 36/37. KIA
Handsford, N.P.L.	LAC	513674	9	Also 36/37
Hanson, H.	AC1	326291	4	
Harding, E.M.	AC1	326675	5	
Harding, G.E.	LAM	7468	1,3	
Hardman, J.D.1.	F/O		4	Also took part In Waz 25 Opns
Hardy, L.	LAC	512436	9	Also 36/37
Hardy, C.A.	AC1	103802	(1)	5 Squadron
Hare, B.W.T.	F/Lt		6	
Hare, C.	Sgt	45733	9	Also 36/37
Hare, S.	Sgt	140	1	
Hare, S.F.	AC1	359114	6	
Harfleet, I.H.	SM1	19786	4	
Hargreaves, W.R.	LAC	370247	6	
Hargroves, J.H.	F/O		5	
Harker, H.J.	Cpl	335599	6	
Harland, W.W.	AC2	296060	3	
Harle, G.	Cpl	511371	9	
Harley, A.	AC1	264645	3	
Harman, J.	Cpl	330142	5	
Harmes, E.W.	LAC	342943	4	

Name	Rank	No.	Clasp	Comments
Harper, H.M.B.	LAC	363751	6	
Harper, W.E.	2/Lt		1	
Harpur, A.J.	AC2	509221	8,9	
Harriman, T.	Cpl	197295	(3)	
Harrington, M.	AC2	299447	1	
Harris, A.C.	Sgt	14698	1	
Harris, A,F.	Lt		1	
Harris, A.T.	S/L		4	
Harris, B.R.	F/O		1,2,3	
Harris, E.H.	AC1	326976	5	
Harris, E.L.	LAC	357817	6	
Harris, F.E.	LAC	506903	6	
Harris, H.C.	LAC	510388	8,9	
Harris, R.	Sgt	67897	6	
Harris, R.C.H.	LAC	362011	6	
Harris DFC AFC, S.B.	F/St		4,6	
Harris, S.E.	LAC	362423	6	
Harrison, G.H.	F/Lt		4	
Harrison, G.H.	F/Sgt	145156	9	Also 36/37 as W/O
Harrison, H.	AC2	329345	3	99 Squadron
Harrison, H.	LAC	408830	1	31 Squadron
Harrison, H.A.	Sgt	335910	6	
Harrison, R.B.	F/O		8	
Harrison, W.A.	Cpl	361760	6	
Hart, A.C.	AC1	349473	4	
Hart, A.R.B.	LAC	330080	5	
Hart, D.R.	LAC	363113	6	
Hart, G.L.	LAC	560279	9	Also 36/37 as Cpl
Hart, J.	Sgt	363975	7	
Hart MC, R.G.	F/Lt		6	
Hart, W.W.	A/W/C		4	
Hartley, C.E.	F/O		6	
Hartwright, W.	AC1	357866	6	
Harvey, A.	F/Lt		8	
Harvey, G.T.	LAC	357664	6	
Harvey, H.F.	F/Lt		9	
Harvey, L.G.	F/O		2,3	20 Squadron
Harvey, S.D.	Sgt	366119	9	Also 36/37
Harvey, T.F.	LAC	356325	6	
Harvie, G.	Cpl	35189	1	
Harwood,	AM2	406155	1	
Harwood, R.A.	LAC	560295	9	Also 36/37
Haslem, J.A.G.	F/Lt		5	
Haslem, R.	AC2	513860	9	Also 36/37 as LAC
Hassall, C.	AC2	345247	4	
Hatch, G.H.	SM2	205169	6	
Hatt, F.J.	AC2	126318	3	
Hatter, S.J.	Cpl	335387	6	
Hattrick, W.H.	AC2	508838	6,8,9	
Havens, P.R.	Sgt	2170	3	
Hawgood, J.V.	AC1	513071	9	Also 36/37 as LAC
Hawke, J.W.	F/O		6	Deceased

Name	Rank	No.	Clasp	Comments
Hawkes, H.C.	Sgt	248206	3	
Hawkes, H.S.	F/O		1	52 Wing
Hawkes, R.H.	LAC	365862	9	Also 36/37
Hawkins, A.E.	AM2	148720	1,2	
Hawkins, C.	LAC	328975	5	
Hawkins, R.E.	Sgt	84631	5	Deceased
Hawkins, W.	Sgt	340821	6	
Hawks, H.C.	Sgt	3729	1	
Hawtin, R.B.	LAC	364834	9	
Hay, C.L.	LAC	363416	6	
Hay, L.R.	AC1	336569	3	20 Squadron
Hay, H.G.	LAC	54908	1,2,3	
Hay, J.B.	Cpl	512376	9	
Hay, W.R.	Capt		5	Army (Political Agent SW)
Haydon, D.A.	LAC	363618	6	
Hayes, H.	Cpl	362474	6	
Hayes, J.C.	A/Cpl	17181	1	
Hayes, P.J.	O/O		3	99 Squadron
Hayles, J.	LAC	343788	4	
Hayman, J.L.T.	LAC	348266	6	
Haynes, F.O.	LAC	330873	4	Also took part in Waz 25 Opns
Haynes, G.	AC2	194767	3	
Haynes, G.R.	LAC	14039	1	
Haynes, P.	F/O		8	Also 36/37 as F/Lt
Hays, J.	AC2	119053	1,2	
Hayter, H.J.	AC1	335575	5	
Hayter, T.H.	Sgt	362949	6	
Hayter-Hames, N.C.	F/O		5	
Haywsrd, E.W.G.G.	AC1	505264	6	
Hayward, F.I.G.	LAC	361994	6	
Hayward, G.A.	LAC	365745	6	
Hayward, R.	F/O		1	
Haywood, A.	F/O		6	
Haywood, F.	AC2	166730	(2),(3)	Application made as Pte in Manchester Regiment
Hazel, F.H.R.	F/Sgt	242991	9	Also 36/37
Hazell, T. F .	S/L		5	
Headford, W.	Cpl	361993	6	
Heal, A.W.J.	LAC	356994	6	
Healy, J.W.	AC1	515261	9	Also 36/37
Healy, C.	LAC	314237	4	
Healy, E.C.	Sgt	115019	6	
Heap, H.	Cpl	341227	6	
Heaps, F.J.R.	Cpl	352148	6	
Heard, C.M.	F/Lt		9	
Heard, G.S.	LAC	510377	8,9	Also 36/37 as Cpl
Heard, W.L.	AC1	108958	1	
Hearsay, G.	AM2	34373	1	
Heasman, M.	AM2	10768	1	
Heath, M.L.	F/O		6	
Hebbert, F.L.B.	F/O		3,4	20 Squadron
Hebden, W.S.	F/O		6	

Name	Rank	No.	Clasp	Comments
Helliwell, F.R.	LAC	505792	6	
Hemingway, W,	Cpl	356267	6	
Hemmings, A.B.	F/Sgt	466	3	20 Squadron
Hemmings, F.A.L.	AC2	507675	6	
Hemsley, B.W.	F/O		6	
Henderson, F.	AC2	338176	6	
Henderson, J.W.B.	LAC	359758	6	
Henderson, R.	O/O		1	
Henderson, S.G.	LAC	370459	6	
Hendley, H.	AC1	342225	4	
Henkers, C. T.	LAC	347598	6	
Henley, W.C.	Sgt	361985	6	
Hennessy, A.		16406S	6	
Henry, R.M.	LAC	342731	4	21/24 in lieu Waz 25
Hensley, G.F.	LAC	506897	6	
Henwood, F.	Cpl	404287	1	31 Squadron
Hepburn, J.	AC2	196176	1	20 Squadron
Hepburn, J.R.D.A.	AC1	327258	4	
Heppell, R.	LAC	359803	6	
Heppenstall, J.E.	LAC	347122	5	
Heppesley, H S.	AM1	407252	1	
Hepple, G.W.	F/Sgt	780	6	
Herbert, F.W.	A/Sgt	355137	9	
Herbert, J.W.	AC2	199297	4	
Heron, D.A.P.	F/O		2,3	
Herriett, A.E.	AC1	321207	4	21/24 in lieu waz 25
Herring, A.F.	AC2	334889	3	
Herrington, C.	Sgt	349451	9	Also 36/37 as F/Sgt
Herriott, G.S.	Sgt	403113	(5)	N.R.V.I.
Heseltine, T.	LAC	560303	9	Also 36/37
Heslop, H.W.	F/O		3,4	
Hester, H.H.C.	Cpl	365363	6	
Hewett, J.D.	Lt		3	
Hewitt, E.N.	Lt		1	31 Squadron
Hewitt, J.M.	LAC	358285	6	
Hewson, W.F.	LAC	511097	9	Also 36/37 as Cpl
Heybourne, H.W.	Cpl	299236	9	Also 36/37
Hickey, J.A.	Sgt	362607	6	Also 36/37
Hicks, J.R.	Cpl	327112	6,9	
Hicks, R.H.F.	LAC	560273	9	Also 36/37
Highton, W.A.	LAC	15520	1	
Hilburn, C.	LAC	363652	6	
Hilder, T.R.	LAC	247385	1,3	20 Squadron
Hill, A. J.	LAC	561945	9	
Hill, C.H.	AC1	353225	5	
Hill, E.L.C.	Cpl	349205	5	
Hill, E.T.H.	Lt		1	
Hill, G.C.	Sgt	48218	1	5 Squadron
Hill, H.F.	AC1	347002	5	
Hill, H.J.	Cpl	351371	6	
Hill, L.C.	AC1	86440	2,3	
Hill, R.	LAC	359571	6	

Name	Rank	No.	Clasp	Comments
Hillman, G.E.	O / O		1	
Hills, F.R.C.	AC2	284853	3	20 Squadron
Hills, G.P.F.	F/O		4	
Hilton, A.	AC1	96710	3	
Hilton, L.	Cpl	343586	4 , 8	
Hilton, S.W.	LAC	350370	6	
Hilton, W.	A/SM	138144	1	
Hind, C.F.	LAC	506792	6	
Hindelang, E.A.	Sgt	363654	6	Deceased
Hindley, J.	AC2	112575	1	
Hine, E.W.	LAC	590249	9	A1so 36/37
Hine, F.A.	LAC	370587	(6)	N.R.V.I.
Hine, J.R.	Sgt	243925	(5)	N.R.V.I.
Hine, L.T.	LAC	359552	6	
Hining, R.	LAC	358179	6	
Hitch, S.W.P.	Cpl	509980	9	
Hitchiner, G.N.	AC1	3S9767	6	
Hitchman, F.	AC1	342272	4	
Hoare, L.I.	AC1	511239	9	
Hoare, L.J.	P/O		1	
Hobbins, A.F.	LAC	159493	5	
Hobbs, J.E.	LAC	159986	6	
Hobbs, O.S.	AC1	334134	2 , 3	
Hockney, J.E.	Cpl	404261	1	
Hodder, C.C.	P / O		9	Also 36/37. KIA. 27 Squadron
Hodges, L.	AM2	408844	1	
Hodgins, W.E.	F/Lt		4	
Hodgson, R.	LAC	328240	4	
Hodgson, S.E.	2/Lt		1	
Hodson, J.W.	LAC	506681	6	
Hoewood, J.	AM3	406081	1	
Hoffling, S.W.F.	LAC	560493	9	
Hoffman, R.G.	LAC	366316	9	Also 36/37
Hogan, F. J.	LAC	364008	6	
Hogan, F.M.	LAC	561561	9	60 Squadron
Hogan, T.E.	Cpl	362518	6	
Hogben, E.T.O'N.	F/O		5	
Hogg, E.C.	LAC	366362	8	
Hogg, R.	LAC	334734	6	
Holburn, W.J.	LAC	345092	5	
Holdaway, S.E.	AC2	288477	4	
Holden, H.E.	Sgt	336576	4	
Holden, R.W.	P/O		5	
Holder, A.F.J.	F/Sgt	180	6	
Holder, L.N.	AC2	560495	6	
Holding, S.C.	LAC	510009	9	
Holdway, H.B.	F/O		5	
Holdway, O.J.	AC2	157781	3	20 Squadron
Holl, C.G.E.	Cpl	49086	6	
Hollamby, A.E.	AC2	65973	1	
Hollamby, W.	AC1	237706	3	97 Squadron
Holland, E.S.	Sgt	59874	(1)	

Name	Rank	No.	Clasp	Comments
Holland, L.W.	AC1	507751	8	
Holland, R.J.	AC1	284779	3	
Holliday, G.J.E.	LAC	243969	4	
Hollinghurst, DFC, L.N.	F/Lt		1,6,8	
Hollingsworth, L.	LAC	508307	9	
Hollis, J.A.	F/Lt		5	
Holloway, A.E.	Sgt	560274	9	Also 36/37
Hollywood, M.	AC2	508033	6	
Holme, H.	Sgt	560770	9	
Holmes, H	Sgt	3094	1	
Honeywell, E.E.	AM1	114101	1,(3)	
Hood, J.	LAC	363103	6	
Hook, C.	LAC	512482	9	
Hook, H.V.	AC1	354472	6	
Hoole, E.T.	Cpl	249274	5	Also entitled to 21/24
Hooper, G.A.	AC1	363076	6	Deceased
Hooper, P.J.	Cpl	349643	9	
Hope, A.	AC2	295424	(1)	5 Squadron
Hope, A.J.E.	LAC	328048	5	
Hope, R.	LAC	342521	5	
Hopkins, E.	AM2	219452	1	Medal O.C.
Hopkins, J.R.	AC2	149684	1	31 Squadron
Hopps, C.	LAC	511569	9	
Horan, E.J.	LAC	352833	6	
Horley, F.A.	Cpl	362985	6	
Horley, F.E.	Lt		1	
Horne, J.T.	Cpl	343722	4	
Horne, K.J.	Sgt	403883	1	
Horner, G.K.	F/O		6	
Horner, L.R.	Cpl	354048	6	
Horrocks, N.O.	AC1	514268	9	Also 36/37
Horton, H.	F/Sgt	348794	6	
Horton, J.F.	Cpl	362013	6	
Hoskins, J.G.	AC2	142612	3	20 Squadron
Hoskins, V.	Sgt	11162	1	
Hossack, A.	AM2	405873	1	
Hotham,	AM2	136806	1	
Hotham, E.A.	LAC	344252	4	Also took part in Waz 25 Opns
Houghton, R.T.B.	F/Lt		4,9	
Houldsworth, A.R.	AC2	342526	3	20 Squadron
House, C.C.	F/O		9	Also 36/37 and 37/39
House, H.H.	AC2	407616	1	
How, J.D.	LAC	365376	6,9	
Howard, A.E.	AC2	18667	1,3	
Howard, A.S.	LAC	364012	6	
Howard, G.E.	F/Sgt	12239	1	
Howe, R.S.	F/O		9	Also 36/37. Deceased
Howell, E.B.	*		5	* British Resident in Kashmir
Howell, S.T.	LAC	36117	6	
Howells, C.	Sgt	330121	6	
Howells, T.A.	LAC	364051	6	
Howlett, A.P.	AM1	408829	1	

Name	Rank	No.	Clasp	Comments
Hoy, C.A.	F/O		3	97 Squadron
Hoy, F.W.J.	AC1	357521	6	
Hoyle, W.C.	LAC	509948	9	Deceased
Hubbard, A.N.W.	AC1	365897	6	
Hubbard, R.C.	LAC	561741	9	Also 36/37
Huckler, L.G.	F/Sgt	241623	4	
Hudleston, E.C.	F/St		9	Also 36/37 as S/L
Hudson, A.	AM2	409200	1	
Hudson, B.J.	LAC	561155	9	Also 36/37
Hudson, F.W.	Sgt	355313	6	
Hudson, J.	LAC	365431	6	Deceased
Hudson, S.	LAC	343376	4	
Hudson, T.B.	AC1	512380	9	Also 36/37 as Cpl
Huggard, G.C.	F/O		1,(3)	31 Squadron.
Hughes, A.C.	LAC	505080	9	
Hughes, A.E.	LAC	356509	6	
Hughes, A.V.	AC2	326560	4	21/24 in lieu Waz 25
Hughes, E.S.	Pte	3596408	9	Border Regiment
Hughes, F.E.	AC1	512996	9	A/Cpl
Hughes, G.V.M.	AC1	103804	(1)	5 Squadron
Hughes, J.F.	LAC	370404	6	
Hughes, T.	AC2	330356	3	20 Squadron
Hughes, W.G.T.	AC1	511611	9	
Hughes-Chamberlain J.L.M.De C.	F/Lt		5	
Hull, F.	Cpl	361987	6	
Hull, G.H.	Cpl	244139	9	Also 36/37
Humberstone, H.C.L.	AC1	326284	4	
Humble, T.	F/O		5	
Hume, C.	Cpl	247845	4	
Humphrey, C.H.	LAC	562152	9	Also 36/37
Humphrey, J.	LAC	560282	9	
Humphreys, F.J.W.	F/Lt		4	
Humphreys, J.Y.	F/O		9	
Humphreys, W.E.	Capt		1	
Hunt, D.S.G.	AC1	564712	9	Also 36/37 as LAC
Hunt, F.H.	AC1	364664	6	
Hunt, G.D.	F/Sgt	2577	1	
Hunt, J.A.	LAC	338018	4	
Hunt, S.	LAC	514306	9	Also 36/37 as AC1
Hunter, E.H.	AC1	370590	6	Also 36/37 as Sgt
Hunter, J.	AC	344869	4	21/24 in lieu Waz 25
Hunter, R.F.	Sgt	365984	9	60 Squadron. Sgt/Pilot
Hunter, R.T.	F/Sgt	333837	6	
Husband, J.H.	LAC	248282	4	21/24 in lieu Waz 25
Huscroft, T.E.	LAC	365379	6	
Hutchieson, J.	F/Lt		6	
Hutchings, K.W.	AC2	340578	4	
Hutchinson, H.	AM2	21140	1	
Hutchinson, J. R.	AC1	513596	9	Also 36/37 as LAC. 60 Squad.
Hutchinson, K.	AC1	512264	9	Also 36/37 as Cpl
Hutchinson, S.W.H.	LAC	364674	8	

Name	Rank	No.	Clasp	Comments
Hutchinson, W.J.	F/O		5	
Hutchison H.H.	Cpl	49392	1	
Hutton, A.F.	F/Lt		9	
Hutton, D.J.C.	F/O		1	31 Squadron
Hutton, H.R	LAC	264545	5	
Huxham, G.H.	F/Lt		8	
Hyatt, T.F.	LAC	560281	9	
Hyde, W.C.	AM2	47164	1	
Ide, C.C.	LAC	335549	(5)	N.R.V.I.
Iden, A.E.	LAC	362015	6	
Ingham, H. F.	AM3	149746	1	
Ingham, J.	AC2	18349	1	
Insley, W.J.	LAC	365422	9	
Ions, G.W.	LAC	370042	6	
Irvine, H.	L/Cpl	3309229	9	Army
Irvine, J.W.	AC	157255	(2),(3)	N.R.V.I.
Irwin, J.C.	AC1	507977	6,8,9	
Isaac DFC, F.H.	F/Lt		6	
Isemonger, A.D.	F/O		9	
Isherwood, F.A.	AM2	149175	1	
Isherwood, W.	AC1	505661	6	
Ivelaw-Chapman, R.	F/O		2,3,4	97 Squadron
Ivey, A.W.	SM1	2132	1,3	
Jackson, A.	AC2	60958	3	
Jackson, A.A.	LAC	355834	6	
Jackson, A.E.	LAC	345651	4	
Jackson, C.	F/O		3,4	99 Squadron
Jackson, C.E.	AC1	250852	1	20 Squadron
Jackson, F.	AC1	563664	9	Also 36/37 as LAC
Jackson , F.W.	AC1	357249	6	
Jackson, G.	AC1	560833	9	
Jackson, J.D.	F/O		3	
Jackson, J.E.	Cpl	313783	3	
Jackson, J.W.	W/O	186726	1	
Jackson, R.A.	AM3	167334	1	
Jackson, T.	AC1	505554	6	
Jacobs, P.M.	Cpl	349832	9	
Jacobsen, G.H.	AC1	515441	9	Also 36/37 as LAC
James, A.F.	F/O		5	
James, G.F.J.	LAC	362458	6	
James, J.D.	A/Cpl	90346	1	
James, W.G.	AM3	21207	1	
James, W.J.	Cpl	335030	6	
Jameson, L.	AC1	513227	9	
Jamieson, J.	LAC	358475	6	
Jamieson, J.R.	Cpl	242657	5	A/Sgt
Janaway, A.J.K.	Sgt	314959	3	
Janes, J.D.	A/Cpl	90346	1	20 Squadron

Name	Rank	No.	Clasp	Comments
Jarman, W.G.R.	F/O		9	Also 36/37 and 37/39
Jarred, R.W.	Sgt	362749	6	
Jarrett, J.H.	LAC	362031	6	
Jarrett, R.	Cpl	340831	3	
Jarvie, J.S.	LAC	346783	5	
Jarvis, A.G.E.	LAC	359512	6	
Jarvis, A. T .	LAC	561764	9	
Jay, N.H.	F/O		3 , 4	99 Squadron
Jay, N.L.	AC1	242033	3	20 Squadron
Jaynes, A.L.J.	AM3	407606	1	
Jeffcoate, P.C.	AC1	359661	6	Deceased
Jefferies, W.C.J.	LAC	335585	4	
Jefferson, A.V.	AC1	505201	6	
Jeffery, G.H.	AC1	327916	4	Also took part in Waz 25 Opns
Jeffrey, T.W.P.	Cpl	344812	6	Deceased
Jeffries, E.C.N.	F/O		3	
Jeffries, S.T.	LAC	347468	4	
Jeffs, A.G.	LAC	511057	9	
Jenkin, J.A.	Cpl	345393	4	
Jenkins, A.E.	Cpl	353363	6	
Jenkins, F.W.	F/Sgt	340975	4	
Jenkins, G.H.G.S.	F/O		6	
Jenkins, H.A.	AC1	337290	3	31 Squadron att. 20 Squadron
Jenkins, J.S.	LAC	511024	9	
Jenner, J.H.	Cpl	344669	4	
Jenner, S.C.	LAC	45633	3	
Jennings, E.F.	AC1	349114	4	
Jennings, L.W.	LAC	364854	6	
Jepson, A	LAC	344645	6	
Jervis, E.	AC1	32808	( )	52 Wing
Jessey, L.W.	Cpl	357108	6	
Jewell, E.W.	Cpl	45562	4	
Jewell, J.	Sgt	348321	6	
Jewell, W.J.	LAC	362461	6	
Jimes, A.W.	Sgt	352497	9	Also 36/37
Jobberns, A.	Cpl	335964	6	
John, W.H.	LAC	511604	9	
Johns, E.	Cpl	335613	6	
Johns, H.V.	AC1	132629	3	99 Squadron
Johnson, A.W.	F/Sgt	19885	3	99 Squadron
Johnson, C.E.	LAC	561771	9	
Johnson, E.	AC1	370874	6	
Johnson, E.H.	AC2	516184	9	Also 36/37 as AC1
Johnson, E.L.	LAC	342266	4	21/24 in lieu Waz 25
Johnson, E.P.	Cpl	5544	(2),(3)	N.R.V.I.
Johnson, F.V.	F/Sgt	401646	3	
Johnson, J.	LAC	358846	6	
Johnson, J.	Cpl	349088	4	Also took part in Waz 25 opns
Johnson, J.C.A.	F/O		6	
Johnson, L.H.	AC1	408398	1	114 Squadron
Johnson, R.A.	LAC	512307	9	Also 36/37
Johnson, T.C.	AC1	347046	4	

Name	Rank	No.	Clasp	Comments
Johnson, W.	AC1	362019	6	
Johnston, G.A.G.	F/O		6	
Johnston, G.T.	Sgt	25472	1	27 Squadron
Johnston, J.H.	AC1	96138	1	
Johnston, R.	LAC	508026	6 , 8	
Johnston, T.R.	Sgt	313789	(1)	20 Squadron
Johnstone, W.	LAC	345280	6	
Joisce, E.W.	LAC	562159	9	Also 36/37. Deceased
Jones, A.	Sgt	78299	(1)	5 Squadron
Jones, A.E.	LAC	347437	4	Entitled as AC2
Jones, A.E.W.	LAC	362018	6	
Jones, A.F.	Cpl	342981	4	
Jones, A.F.T.	Cpl	358096	6	
Jones, A.J.	LAC	345362	5	
Jones, B.	AC1	508536	9	Also 36/37 as LAC
Jones, B.P.	F/O		4	21/24 in lieu Waz 25
Jones, C.H.	Cpl	362020	6	
Jones, C.W.	LAC	356953	6	
Jones, D.W.	LAC	354544	6	
Jones, E.C.J.	AC1	512262	9	
Jones, E.M.	LAC	34154	(1)	31 Squadron
Jones, F.	LAC	362017	6	
Jones, F.C.	AC1	329433	3	31 Squadron att. 20 Squadron
Jones, F.G.	Cpl	241197	3 , 4	
Jones, F.H.	Cpl	365374	9	Also 36/37
Jones, G.	LAC	506790	6	
Jones, G.	SM2	134587	6	
Jones, G.E.	AC2	155889	1	
Jones, H.	Sgt	333313	6	
Jones, H.	A/Cpl	370838	9	
Jones, H.E.	AC2	329928	3	
Jones, H.K.	WO2	340008	9	
Jones, H.L.	LAC	363657	6	
Jones, J.	F/Sgt	6768	1	31 Squadron
Jones, J.	AC2	328463	3	
Jones, J.H.	Cpl	330149	6 , 9	
Jones, J.R.	F/Lt		6	
Jones, M.	AC1	356535	6	
Jones, N.C.	F/O		9	
Jones, R.D.	LAC	370777	6	
Jones, R.J.	LAC	363114	6	
Jones, R.O.	F/Lt		9	
Jones, S.C.P.	LAC	363154	6	
Jones, S.F.	LAC	342364	5	
Jones, T.	LAC	370673	6	
Jones, T.	LAC	96775	(1)	5 Squadron
Jones, T.E.	AC1	343652	5	
Jones, T.R.	Cpl	590120	9	Also 36/37
Jones, Y.C.L.	LAC	405078	3	97 Squadron
Jones, W.	AC2	349539	4	
Jones, W.H.	Cpl	404265	1 , 3	
Jones, W.H.	Sgt	361779	6	

Name	Rank	No.	Clasp	Comments
Jones, W.M.	AC2	344008	(1)	5 Squadron
Jopling, W.	LAC	365249	6	
Jordan, C.H.	AC2	343104	4	21/24 in lieu Waz 25
Jordan, L.W.	Cpl	357821	9	Also 36/37 as Sgt
Jordan, R. B .	F/Lt		6	
Josiah, P.	Cpl	248798	3	
Josiffe, C.L.	LAC	363163	6	
Jowitt, L.	LAC	562160	8	
Jowsey, L.S.	Sgt	346779	6	
Joyce, F.	Cpl	342929	4	
Judd, H.L.	AC1	238510	3	20 Squadron
Judd, J.	AC1	342548	4	
Judge, A.	SM2	166040	4	
Judge, G.H.	LAC	590099	8	
Judson, E.H.	Cpl	6166	1	
Judson, W.E.	Cpl	I59941	8	
Jukes, A.S.C.	Sgt	364118	6	
Junor DFC, H.R.	P/O		1,3,4	31 Squadron
Jupp, C.	LAC	343846	5	
Kavanagh, W.	LAC	342417	4	
Kay, T.	LAC	343291	4	
Kearley, C.W.	LAC	510730	8,9	
Keary, C.R.	F/Lt		4	
Keddie, W.M.	F/O		8	
Keeble, E.C.	AC1	347850	6	
Keen, G.F.S.	LAC	370630	6	
Keen, H.	AC1	126315	3	20 Squadron
Keen, H.	LAC	509975	9	Also 36/37 as Cpl
Keeping, E.G.	F/O		1	31 Squadron. Medal O.C.
Keey, M.W.	F/O		6	
Keirl, F.H.	Cpl	155057	6,8,9	
Kelly, A.P.	LAC	366327	8	
Kelly, A.S.	AC2	250889	(1)	20 Squadron
Kelly, J.B.	LAC	342729	5	
Kelly, M.R.	F/O		(6)	Forfeiture
Kelly, T.P.	LAC	353003	6	
Kelly, W.	AC2	337848	3	20 Squadron
Kelly, W.J.	LAC	247939	2,3	
Kemp, F.E.	AC1	161770	3	20 Squadron
Kemp, S.C.	AM1	13868	1,2,3	
Kempster, R.	LAC	561609	9	Also 36/37
Kempthorn, F.	AC1	343733	5	
Kenchington, S.	AC2	403174	1	
Kendall, J.G.	LAC	358082	6,8,9	
Kennedy, A.J.	F/O		9	
Kennedy, J.E.	Sgt	352849	6	
Kennedy, O.A.	Lt		4	Issued as Lt R.A.
Kennedy, R.	AC1	195330	3	
Kennedy, W.	Cpl	345451	4	
Kenny, F.	Cpl	46802	3	20 Squadron

Name	Rank	No.	Clasp	Comments
Kent, P.F.J.	Capt		4	
Kerens, F.	AC1	561572	9	
Kerr, A.W.C.	AC1	563826	9	Also 36/37
Kerrison,	AM2	39338	1	
Kerry, F.C.	Cpl	347331	4	
Kesterton, H.P.M.	F/Lt		1,2,3	
Kettle, F.A.	LAM	25106	1	
Kettley, J.R.	Sgt	157434	6	
Kewell, W.R.	AC2	340840	3	
Kidson, A.H.	LAC	505090	6	
Kilbride, F.J.	LAC	358552	6	
Kilgour, C.G.	LAC	590301	9	
Kilpatrick, R.E.	Cpl	330789	5	N.R.V.I.
Kilsby, T.H.J.	AM2	60543	1	
Kilvington, J. C.	AC2	149220	( )	20 Squadron
Kimber, R.E.	Cpl	561570	9	
Kinch, G.W.	LAC	358127	6	
King, E.	Sgt	9972	1	
King, G.R.	LAC	512179	9	Also 36/37
King, H.E.	F/Lt		1,2,3	
King, L.E.	Cpl	364024	7	
King, W.T.	P/O		9	Also 36/37. Deceased
King-Lewis, A.	F/O		5	
Kings, F.F.	LAC	346277	4	
Kingsland, W.H.	Lt		1	
Kingston, L.F.	F/Sgt	401704	2,3	20 Squadron
Kinloch, H.	LAC	365495	6	
Kinnear, D.	Sgt	335127	5	
Kinsey, J.A.	AC1	505919	6	
Kirby, F.J.	LAC	561776	9	Also 36/37
Kirby, J.L.	F/O		1,2,3	31 Squadron
Kirk OBE, J.E.	P/O		9	Also 36/37 and 37/39
Kirkham, D.	Sgt	91026	6	
Kirkham, N.	P/O		6	
Kirkham, W.	AC2	156289	3	20 Squadron
Kitching, F.J.	AM2	300809	1	
Klyne, D.D.	Sgt	358618	6	
Knight, A.S.	AC1	342302	4	
Knight, L	AM2	84891	1	
Knight, R.P.G.	Cpl	49212	( )	20 Squadron
Knocker, G.M.	F/O		2,3,4	
Knowles, J.	LAC	505239	6	
Knowles, J.S.	LAC	563804	9	Also 36/37
Knowlton, H.G.	AC1	5187	3	
Koch, C.T.D.	Cpl	331044	4	21/24 in lieu Waz 25
Lade, T.J.	ACl	364111	8	
Ladell, A.E.W.	ACl	341611	4	
Laidler, H.E.	Sgt	340396	9	
Laing, D.	ACl	506188	6,8,9	
Lainson, L.H.	AC1	50214	3	20 Squadron

Name	Rank	No.	Clasp	Comments
Lale, H.P.	F/Lt		1,2,3	
Lamb, F.J.	Cpl	362945	6	
Lamb, J.F.	AM2	14723	1	
Lambert, H.A.	AC2	249629	3	
Laming, M.R.	Sgt	335330	6	
Lancaster, A.L.	LAC	342611	5	
Lancaster, J.	LAC	246849	3	97 Squadron
Lancaster, J.	LAC	349455	4	Also took part in Waz 25 Opns
Lane, A.	F/Sgt	221859	1	
Lane, D.W.	F/O		6	
Lane, H.	Sgt	363639	6	
Lang, A.	LAC	335966	6	
Lang, S.	LAC	561779	9	Also 36/37 as Cpl
Lang, T.	AC1	407246	I	
Langdown, R.F.	AC2	327682	5	
Langford, A.S.	W/O	26529	(3)	
Langhorne, F.R.	LAC	363167	6	
Langley, C.D.	LAC	364902	6	
Langley, C.F.	LAC	337630	3	99 Squadron
Lansdell, W.H.	LAC	335828	6	Deceased
Lapraik, D.M.	F/O		1	20 Squadron
Lapthorn, J.F.	LAC	365284	9	Also 36/37 as Cpl
Larner, W.C.	AC2	507752	6	Also 36/37 as Cpl
Lart, E.C. de V.	F/Lt		6	
Laslett, M. H.	AC1	313156	4	
Lathe, E.C.	AC1	506815	6	
Laurence, F.H.	F/Lt		5	
Lavender, L.A.	F/Lt		3	20 Squadron
Laves, G.	AM1	133068	1	
Law, J.B.	SM2	338068	(1)	N.R.V.I.
Law, J.R.	Cpl	327385	4	
Lawe, A.G.	F/O		1,2,3	20 Squadron. Officer returned to Army duties
Lawler, W.J.	AC1	343016	4	21/24 in lieu Waz 25
Lawless, E.J.	AC1	296415	4	
Lawrance, J.J.	LAC	512285	9	
Lawrence, A.G.	AC2	334023	3	20 Squadron
Lawrence, E.G.	AC1	343898	4	21/24 in lieu Waz 25
Lawrence, G.	Cpl	23696	1	
Lawrence, J.	AC2	329242	2,3	
Lawson, E.	F/O		1	
Lawton, H.	AC2	293772	3	20 Squadron
Lawton, T.	AC1	332175	3	31 Squadron att. 20 Squadron
Laybourne, E.C.	Cpl	335590	6	
Layden, J.	Cpl	341674	4	
Lazzary, G.E.	AC1	513228	9	Also 36/37 as LAC
Lea, H.S.	AC1	330418	4	Also took part in Waz 25 Opns
Lea, S.L.	LAC	342968	4	21/24 in lieu Waz 25
Lea-Cox, C.L.	F/Lt		6,8	
Leader,	AM2	171994	1	
Leahy, J.D.	F/Lt		4	
Leake, A.G.	AM2	115244	1	

Name	Rank	No.	Clasp	Comments
Leaver, W.C.	LAC	350176	4	21/24 in lieu Waz 25
Leavesley, W.G.H.	LAC	363179	6	
Leclere, H.E.	LAC	509794	9	
Ledbury, W.	AC2	269S27	3	
Ledger, H.	Cpl	366142	6	
Lee, H.W.	LAC	352675	6	
Lee, D.J.P.	F/O		9	
Lee, S.E.	AC1	351442	5	
Lee, W.A.	F/O		9	Also 36/37 as F/Lt
Leech, L.G.	Cpl	362070	6	
Leeden, G.H.	F/Sgt	28026	1	
Lees, AFC, A.	W/C		9	A1so 36/37 as G/C
Leese, C.H.	Cpl	356165	6	
Leese, G.	AM2	408834	1	
Le Gallienne, C.E.	LAC	49199	2,3	
Legg, E.G.	LAC	362950	6	Also 36/37 as F/Sgt
Le Good, E.J.	Cpl	362079	6	
Leonard, J.A.	F/O		1	Medal O.C.
Leopold, E.	F/Sgt	22	4	
Le Poer-Trench, C.F.	F/Lt		6	
Lepper, J.H.	AC1	364124	6,8	
Leslie, A.W.	AC1	561792	9	Also 36/37 as LAC
Leslie, J.S.	F/O		9	Also 36/37 as F/Lt
Lester, C.W.	LAC	249279	3	20 Squadron
Lester, H.L.P.	F/O		(5)	N.R.V.I.
Lett, F.	Sgt	108963	3	97 Squadron
Levers, D.	Cpl	5196	1	
Lewer, L.H.	LAC	511071	9	
Lewin, B.A.S.	F/O		2,3,4	99 Squadron
Lewis, H.F.A.	AC1	359200	6	
Lewis, A.T.	AC2	329499	3	20 Squadron
Lewis, C.B.	AC1	515625	9	
Lewis, D.H.	Cpl	33922	1	
Lewis, E.	LAC	334311	3	20 Squadron
Lewis, E.G.	LAC	39185	2,3	
Lewis, G.A.	LAC	561781	9	
Lewis, G.W.G.	Cpl	356393	9	
Lewis, J.	F/Sgt	685	1	
Lewis, J.G.	Cpl	362477	6	
Lewis, J.W.	Cpl	340264	6	
Lewis, L.J.	Sgt	245906	3	
Lewis, L.R.	Cpl	343808	4	
Lewis, S.	AC2	404260	1	
Lewis, T.G.	LAC	356185	6	
Lewis, W.F.	Sgt	4311	1	
Lewis, W.R.	LAC	366126	9	Also 36/37 as Cpl
Liddell, T.	LAC	309791	3	
Liddle, W.	AC2	328245	3	20 Squadron
Lillie, H.C.	AC2	59482	1	
Lincoln, C.	LAC	507872	6	
Lindfield, E.	AC1	84624	3	20 Squadron
Lindley, W.J.H.	F/Lt		9	

Name	Rank	No.	Clasp	Comments
Lindop, V.S.E.	F/Lt		3,4	28 Squadron att. 20 Squadron
Lindsay, J.	LAC	354634	6	
Lindup, C.A.	F/Lt		5	
Lines, G.F.	F/O		4	
Lines, S.R.	AC1	357175	6	
Lingard, H.	AC1	560314	9	Also 36/37 as LAC
Lipscomb, S.W.	LAC	506148	6,8	
Lipscombe, R.P.	Cpl	239933	6	
Lissett, J.W.	F/Lt		6	
Lister, R.C.F.	P/O		9	Also 36/37 as F/Lt
Little, F.	Cpl	590086	9	
Little, J.	AC1	508431	6	
Little, W.E.	Sgt	349408	9	
Littleproud, C.	F/Sgt	1014	6	
Littlewood, C.E.	LAC	329342	3	Entitled to Mahsud.Not claimed. 99 Squadron
Livermore, C.	Cpl	86350	1	5 Squadron
Livock DFC AFC, G.E.	S/L		7	
Llewellyn, C.F.	Cpl	364130	6	Also 37/39 as F/Sgt
Llewellyn, L.C.	Cpl	335823	6	
Lloyd, A.R.	LAC	343350	5	
Lloyd, C.R.D.I	F/O		9	
Lloyd, H.P.	F/Lt		4,8,9	
Lloyd, J.H.	LAC	157462	(4)	N.R.V.I.
Lloyd, W.E.	LAC	362052	(6)	
Loader, H.G.	AC1	408847	1	31 Squadron
Loader, W.J.C.	Sgt	352517	9	Also 36/37 as F/Sgt
Loades, F.G.	Cpl	336304	6,9	
Lobley, J.O.H.	F/O		6	
Lochead, D.G.	Sgt	240302	6	
Lockett, A.	Cpl	349121	9	Also 36/37 as A/Sgt
Lodder, H.W.C.	AC1	245249	4	
Lofthouse, G.	AM1	12700	1	
Logan, R.H.	LAC	366241	9	
London, H.D.	AC1	63808	1	
Lonergan, H.R.F.	LAC	506917	6	
Long, H.E.	Sgt	406089	1	31 Squadron
Long, L.	AM	13123	1	
Long, T.	LAC	507449	9	
Longbottom, G.G.	AM2	408841	1	
Longfellow, G.D.	F/O		1	31 Squadron Deceased
Longman, A.	AC1	333624	3	20 Squadron
Longman, C.J.	Cpl	355028	9	
Loosemore, C.A.	AC1	59860	1,2,3,4	
Lord, D.G.R.	F/O		1	48 Squadron
Lord, V.L.A.	AC2	149186	3	20 Squadron
Loughnan, C.H.	F/Lt		6	
Love, R.G.	AC1	363i98	6	
Lovegrove, J.C.S.	LAC	363173	6	
Lovell, A.J.	LAC	365497	6	
Lovell, F.S.	Sgt	340964	6	
Lovett, F.	Cpl	342871	4	

Name	Rank	No.	Clasp	Comments
Lovett, L.H.	Cpl	357589	9	
Low, J.H.	P/O		1	
Lowe, F.	Sgt	331499	6	
Lowe, H.E.	LAC	350457	4	
Lowe, S.	LAC	506848	6	
Lowe-Holmes, P.W.	F/O		6	
Lowry, H.R.	F/O		6	
Luard, R.B.	F/O		4	Deceased
Lucas, S.G.	Sgt	342385	6	
Lucas, W,C.	LAC	366339	9	
Lucas, W.S	AC2	509571	( )	N.R.V.I.
Luck, H.F.	F/O		4	
Luck, R.E.	AC1	562181	9	
Lockhurst, J.E.	AC1	84069	1 , 3	
Lucy, R.F.	Cpl	365191	6	
Luke, W.	AC2	340814	3	52 Wing
Lund, A J.	LAC	362053	6	
Lund, E.L.	AM2	187390	1	
Lunn, N.A.C.	LAC	362479	6	
Lusk, P.B.	F/O		6	
Lusty, A.W.	LAC	365882	6	Deceased
Lynch, J.	F/Sgt	343656	5	Medal issued 23.1.46
Lyon, H.W.G.	LAC	356567	6	
Lyons, T.F.	Sgt	294394	4	
Lyster, C.H.E.	F/O		9	Also 36/37
Mabey, L.K.	LAC	348886	4	
McAllister A.B.	LAC	995	2,3	A/Sgt
McAllister, F.G.	Sgt	200697	1 , 4	
Macandie, A.	AC1	20077	1	31 Squadron
Macario, A.	LAC	357073	(6)	N.R.V.I.
Macario, V.G.	Sgt	205786	1 , 3	31 squadron
McBain, J.	F/Lt		6	
McBride, W.	Sgt	359753	9	
McCaffray, J.C.	AC2	328132	4	
McCarthy, H.C.	AC1	505802	6	
McCay, H.	AC1	505347	6	
McClurkin, T.	S/L		1	War Office issue
McCormack, G.	O/O		1,2,3	
McCudden, J.A.	LAC	560344	9	Also 36/37 as Cpl
McDonald, A.	LAC	366354	8,9	
MacDonald, A.H.H.	F/Lt		6	
MacDonald, D.	AC1	359300	6	
MacDonald, D.	AC2	332212	3	99 Squadron
MacDonald, F.M.	Sgt	7800	( )	
McDonald, H.	F/Sgt	313815	1	20 Squadron
MacDonald, J.	A/Sgt	346843	4	
MacDonald, K.	AC2	190646	1	5 Squadron
MacDonald, W.	AC2	336799	4	
McEleney, J.	LAC	344958	4	
McEwan, D.	LAC	364195	8,9	

Name	Rank	No.	Clasp	Comments
McEwen, N.D.K.	W/C		1	Medal O.C.
Macey, F.	LAC	370846	6	
Macey, P.O.	Cpl	362098	(6)	N.R.V.I.
Macey, W.	F/O		9	
MacFarlane, W.	Sgt	3097	1	
McGowan, W.	O/O		1,2,3	20 Squadron
McGowan, W.A.	LAC	561865	9	Also 36/37
McGowen, E.	F/O		1	
MacGregor, C.S.	F/O		1	
MacGregor, T.A.	Sgt	1265	1	
McGuire, J.	LAC	245478	4	
McGunigal, J.S.	AC1	345276	4	
MacIver, J.	AM1	75654	1	
Mack, W.	AC1	508683	6,8	
Mackay, G.F.	F/O		4	
McKay, J.	F/Sgt	266742	6	
Mackay, M.B.	F/Lt		9	
McKeand, G.	AC1	508564	6,8	
McKee, A.	F/O		6	
McKeever, S.	F/O		4,6	
McRenzie, J.W.	LAC	343276	6	
Mackie, J.McC.	Cpl	133605	6	
McKinley, J.P.J.	LAC	562572	9	Also 36/37
McKinley-Hay, E.A.	F/Lt		6	
McKinney, J.	AC2	513673	9	Also 36/37 as AC1
Mackinnon, A.A.	Cpl	505082	6	
McLean, C.P.	Cpl	330131	4	
McLellan, J.A.	LAC	505590	6	
McLellan, S.	AC1	248049	1	20 Squadron
McLeod, A.N.	AC1	343158	3	20 Squadron
McLeod, W.J.	LAC	344238	(5)	N.R.V.I.
McLoughlin, A.	AC1	562790	8,9	Also 36/37 as LAC
McManus, E.S.	LAC	363176	6	
McManus, W.J.J.	AC1	285000	4	
McMillan, A.R.	Cpl	334786	4	21/24 in lieu Waz 25
McMorran, J.	Sgt	157262	6	
McNab, A.	Sgt	18134	4	
McNally, C.H.R.	LAC	364892	6	
McNeil, H	LAC	326587	5	
McNeil, H.	LAC	362405	(6)	N.R.V.I.
McNesby, E.G.	LAC	343268	4	
MacPherson, R.C.	Capt		4	R. Highlanders seconded toRAF
McQueen, J.	AC2	156864	1	
McShane, J.C.	AC1	77766	1	
McWilliam, A.	Cpl	347781	6	
Madders, C.T.	LAC	355240	6	
Maddox, J.H.	LAC	332289	3	20 Squadron
Maddox, R.H.	LAC	359928	6	
Madgwick, W.R.S.	AC2	512630	8,9	
Maddison, F.	AC2	138159	1	
Maggs, R.P.	AC1	359199	6	
Magner, J.	F/Lt		6	

Name	Rank	No.	Clasp	Comments
Maher, J.E.	AC2	343798	(5)	N.R.V.I.
Mahon, J.	AC2	337000	3	20 Squadron
Mahony-Whitton, G. W.	F/O		3	52 Wing
Maidment, E.	Cpl	590100	9	
Maidment, E.W.	AM1	7513	1	
Mair, A.	AC1	147133	5	
Major, M.C.J.	LAC	365277	6	
Mallandain, W.L.	AC2	131158	1,3	
Mallinson, L.V.	Cpl	356835	8	
Maltby, H.	AM1	14314	1	
Maltby DS0 AFC, P.C.	S/L		4	
Malyan, B.J.	O/O		1,2,3	27 Squadron
Malyon, F.C.	LAC	358735	6	
Mangan, W.	LAC	357397	6	
Mangles, R.A.R.	F/O		5	
Manklow, A.E.	AC1	326645	4	21/24 in lieu Waz 25
Manley, F.H.	A/Sgt	239764	4	
Mann, A.J.	LAC	511599	9	
Manning, E.	AM1	23727	1,2,3	
Manning, F.	LAC	345547	5	
Manning-Fox, J.H.	F/O		6,8	
Manser, F.A.	LAC	341916	4	
Mansfield, H.L.	AC1	250988	3	
Manson, W.A.	2/Lt		1	
Mantle, T.F.	F/Sgt	313171	2,3	20 Squadron
Marchant, S.	AC2	187864	1	
Marett, P.L.P.	F/Lt		9	Also 36/37 and 37/39
Markall, L.	LAC	347674	4	
Marketis, J. B.	LAC	361774	6	
Marks, J.	LAC	342531	4	21/24 in lieu Waz 25
Marks, R.	LAC	314294	4	
Marland, F.	LAC	514450	9	
Marriott, E.	ACl	157336	4	
Marsh, E.A.	AC1	515130	9	Also 36/37 as LAC
Marsh, F.H.	Cpl	364197	6	
Marsh, H.A.D.	AC1	347675	5	
Marsh, K.J.P.	AC1	516155	9	Also 36/37 as LAC
Marsh, S.V.	AC2	78162	3	
Marsh, S.W.	AC1	349142	5	
Marshall, C.R.	F/Sgt	89800	9	Retd to Mint for scrapping. Also 36/37. 60 Squadron
Marshall, F.H.	AC1	565640	9	Also 36/17 as LAC
Marshall, H.	AC2	507263	6	
Marshall, L.H.	LAC	509148	9	
Marsland, F.	AM2	303140	1	
Martin, A.E.	AM2	407608	1	
Martin, H.G.	LAC	365104	6,9	
Martin, H.S.	F/O		6	
Martin, J.	AC1	341754	4	
Martin, J.	AC1	348469	5	
Martin, J.W.	AC2	129680	3	
Martin, W.	AC1	334537	3	20 Squadron

Name	Rank	No.	Clasp	Comments
Martindale, C.	LAC	514648	9	Also 36/37
Marwood, H.	LAC	358583	6	
Mason, G.	LAC	561274	8,9	
Mason, J.S.	Cpl	343873	5	A/Sgt
Mason, K.V.	Cpl	365496	6	
Mason, N.W.F.	P/O		5	
Massy, D.F.	F/Lt		1	31 Squadron
Masters, C.H.	F/Lt		4	
Matheson, A.	AC2	332837	3	20 Squadron
Mathews, J.R.	F/O		6	
Mathews, S.	AM2	408328	1	
Matley, T.W.	AC2	331355	3	
Matthews, E.	LAC	350366	6	
Matthews, E.G.	AC2	140364	1	20 Squadron
Matthews, F.	Cpl	365893	9	
Matthews, H.G.	LAC	362081	6	
Matthews, H.G.G.	F/Sgt	246459	9	Also 36/37 as W/O
Matthews, R.L.	AC1	148721	3	52 Wing
Mauchel, W.E.	F/Sgt	14008	1	
Maud, C.C.	AC1	370555	6	
Maude, W.A.	Cpl	327410	5	A/Sgt
Maule, E.H.	Cpl	560365	8,9	
Maund, A.C.	W/C		4	
Maver, G.A.	AC1	246086	4	
Mawdsley, H.E.J.	LAC	364141	9	Also 36/37 as Cpl
May, W.H.	Cpl	239011	8,9	
Mayell, S.B.	LAC	202054	4	21/24 in lieu Waz 25
Mayhew, S.F.	LAC	355603	3,4	
Maynard, J.W.	LAC	510398	9	
Mayne, E.	Sgt	82478	6	
Mayo, H.H.	AC2	59586	1,2,3	
Mead, G.E.	AC2	509995	6,9	
Mead, J.F.	Lt		1	
Meagor, S.J.	Cpl	362065	6	
Meaker, T.	AC1	245911	1	31 Squadron
Meally, W.H.	AC1	511624	9	
Meares, G.W.R.	AC	365898	6	
Meech, R.J.	LAC	347517	5	
Mein, F.		342414	4	21/24 in lieu Waz 25
Melitus, P.N.	F/Lt		4	
Melvin, L.A.	LAC	359855	6	
Mendoza, E.S.	AC2	243403	1	
Menzies, R.	F/O		6,8	
Mercer, W.K.	F/O		3,6	5 Squadron
Meredith, D.	LAC	332369	5,6	
Meredith, H.	AM1	33772	1	
Meredith, J.H.	AC2	349462	4	
Merer, J.W.F.	F/O		1,2,3,4	
Merron, G.D.	WO2	401495	9	
Metcalfe, E.W.	Cpl	326834	4	21/24 in lieu Waz 25
Metcalfe, R.	AC1	358752	6	
Metcalfe, S.	AC1	505462	6	

Name	Rank	No.	Clasp	Comments
Metcalfe, S.M.	AC1	509060	9	
Meyer, V.J.H.	Cpl	560666	9	
Mickels, G.T.	AC1	363678	6	
Middlemist, W.	AC2	511815	6,9	
Middleton, L.G.	LAC	357916	6	
Middleton, W.	LAC	359705	7	
Milburn, H.D.	AC2	158082	1	20 Squadron
Miles, A.G.	LAC	560847	9	
Miles, J.E.	LAC	508812	9	
Miles, R.	AM2	6316	1	
Mill, M.F.G.	F/Sgt	238686	4	
Millar, C.S.	2/Lt		2,3	
Millar, E.L.	S/L		(1)	31 Squadron
Millar, R.D.	AC2	332201	3	20 Squadron
Miller, A.	Cpl	32045	1	
Miller, A.E.	Sgt	245551	6,9	
Miller, A.J.	Cpl	246089	6	
Miller, E.C.	LAC	514062	9	Also 36/37
Miller, F.	LAC	364705	6	
Miller, G.	AM2	102809	1	
Miller, G.I.	LAC	370111	6	
Miller, H.S.	AC1	149898	4	21/24 in lieu Waz 25
Miller, J.W.	LAC	342817	5	
Millhouse, C.A.	T/Cpl	222637	(1)	20 Squadron
Millman, J.S.	LAC	561821	9	Also 36/37 as Cpl
Mills, A.J.	LAC	341800	4	
Millyard, S.	AC2	251458	1,2,3	
Milne, J.R.	AC1	341810	4	
Milsom, V.R.	LAC	347859	5	
Minchin, F.F.	W/C		1	
Minns, G.W.	AC1	359262	6	
Mitchell, A.	AC1	506589	6	
Mitchell, C.E.	Cpl	364132	6,8	
Mitchell, D.R.	F/O		4	
Mitchell, F.F.G.	Cpl	38827	1	
Mitchell, F.G.S.	F/O		5	
Mitchell, H.M.	LAC	362059	6	
Mitchell, J.F.	AC1	327389	5	
Mitchell, J.H.	AC1	370484	6	
Mitchell, M.P	LAC	562809	9	
Mitchell, N.	LAC	359366	6	
Mitchell, P.J.	F/Sgt	1447	6	
Mitchell, T.N.	Cpl	343239	4	
Mitchell, W.G.S.	W/C		3,4	52 Wing
Mitchener, A.	AC2	247869	1,3	20 Squadron
Mitchinson, S.H.	LAC	513922	9	Also 36/37 as Cpl
Moat, A.S.	A/Sgt	335366	9	Also 36/37 as Sgt
Moffatt, A.B.	LAC	330678	4	
Moger, J.H.	AC2	328520	3	20 Squadron
Moir, D.A.	AC2	506225	6	
Moir, J.R.	Sgt	364949	6	
Moland, H.J.	LAC	328571	4	

Name	Rank	No.	Clasp	Comments
Molesworth, A.O.	F/O		6	
Mollard, R,P.	F/O		4	
Mollison, J.H.A.	P/O		5	
Molyneux, R.	LAC	560364	9	
Monk, G.W.	F/O		6	
Monks, A.T.	Sgt	364944	6	
Monro-Higgs, W.R.	F/O		8	
Montgomery, G.	Cpl	307222	9	Also 36/37
Montgomery, J.	LAC	511346	9	
Moody, H.M.	F/Lt		4	
Moody, J.	AC2	342247	4	
Moon, F.J.	LAC	362055	6	
Moore, A.H.	AC2	509399	3	Engraved Pte Norfolk Regiment
Moore, F.	Cpl	15338	1	
Moore, F.J.B.	Cpl	364144	6	
Moore, H.W.S.	Cpl	353099	6	
Moore, J.	AC1	505596	6	
Moore, M.	F/O		2,3	31 Squadron att. 20 Squadron
Moore, S.J.	AC1	331556	3	20 Squadron
Moore, T.	AC1	407540	3	
Moore, W.I.	AC1	347001	4	
Morcom, F.C.	AC1	249224	1	48 Squadron
Moreton, N.V.	F/O		4	
Morey, J.W,G.	AC1	334699	3	99 Squadron
Morgan, A.J.	LAC	365505	6	
Morgan, F.	AC1	511388	8	
Morgan, J.C.	AC2	350070	5	
Morgan, L.	Sgt	245780	5	
Morgan, W.	WO2	332470	8	Not entitled to NWF 1935
Morley, F.H.	AC2	251033	3	99 Squadron
Morley, J.K.	AC2	510419	6	
Morman, W.T.C.	Sgt	560125	9	Also 36/37 as Sgt. Deceased
Morrell, S.J.	F/O		1	'Aden Flight'
Morrin, R.A.	LAC	348569	4	21/24 in lieu Waz 25
Morris, A.P.R.	AC1	278248	3	
Morris, B.	AC2	125715	3	
Morris, E.	AC1	341838	(5)	N.R.V.I.
Morris, G.	AC1	363677	6,8	
Morris, G.S.	Sgt	362076	6	
Morris, H.	LAC	510795	9	
Morris, J.H.	Cpl	329387	3	20 Squadron
Morris, R.R.G.	Cpl	335700	6	Also 36/37 as Sgt
Morris, S.E.	LAC	106613	6	
Morris, W.	A/Cpl	16911	1,2,3	
Morris, W.A.	AC1	515406	9	Also 36/37 as Cpl
Morrish, F.T.	AC2	263415	1	
Morrish, W.R.	LAC	342126	4	
Morriss, I.C.	LAC	50073	1	
Morse, W.C.	LAC	511731	9	
Mortimer, E.E.	LAC	505527	6	
Mortimer, R.	AC1		1	
Morton,	AM1	35878	1	

Name	Rank	No.	Clasp	Comments
Morton, V.	LAC	513440	8,9	Also 36/37 as Cpl
Moss, E.G.	LAC	363662	6	
Moss, F.	Cpl	509120	9	Also 36/37
Moss, S.G.G.	AC1	363477	6	
Moss, W.E.	Cpl	3462S6	6	A/Sgt
Mote, P.J.	F/O		6	
Motley, C.S.	LAC	511915	9	
Mottram, A.E.	AC1	84948	3	
Mouland, T.C.	LAC	590233	9	
Mounsey, D.	AC2	333863	2,3	99 Squadron
Mounsey, R.J.	S/L		3	
Mounter, B.	LAC	352872	6	
Moxey, R.V.	Sgt	365105	6	
Moyes, C.T.M.S.	LAC	506602	6	
Moyles, S.	P/O		1	
Muir, C.	AC1	340953	4	
Mulford, J.A.	LAC	370350	6,8	
Mullan, D.R.	F/O		1,2,3	20 Squadron. Medal engraved 'D. Mullan'. Deceased
Mulley, J.	LAC	347442	4	
Mullinger, R.0.	P/O		1	
Mumby, C.N.A.B.	P/O		5	
Mummery, A.E.	LAC	349352	4	
Mummery, W.H.	AC1	202455	1	
Munday, C.W.	Cpl	335506	6	
Munday, V.G.	A/Cpl	59655	1	31 Squadron
Mundell, W.F.	AC1	347434	(5)	N.R.V.I.
Munden, J.E.	Cpl	335965	6	
Munro, D.J.	Sgt	337793	5	
Munson, E.G.	O/O		1	
Murden, A.H	LAC	364912	9	
Murgatroyd, P.	F/O		1	31 Squadron
Murison, F.W.	F/O		6	
Murphy, J.P.	AC2	365506	6,8,9	
Murphy, L.	F/O		2,3	
Murphy, P.E.	Cpl	342680	4	21/24 in lieu Waz 25
Murphy, R.	LAC	563571	9	Also 36/37
Murphy, W.	Sgt	333569	3	20 Squadron
Murphy, W.E.	LAC	335987	6	
Murray, D.G.	AC1	364710	6	
Murray, E.M.	W/C		4	
Murray, G.	AC1	511736	9	Also 36/37
Murray, J.P.	AC1	513996	9	Also 36/37
Murray, J.R.C.	LAC	248348	(1)	5 Squadron
Murray, J.S.	AC2	285434	3	
Murray, P.	AC2	344856	5	
Murray, W.B.	F/O		9	
Murrell, A.E.	AC1	370399	6	
Musselwhite, C.C.	F/O		5	Waz 25 in lieu 21/24
Muzeen, R.	LAC	505028	6	
Mylne, A.W.	S/L		4	

Name	Rank	No.	Clasp	Comments
Nagle, J.G.	P/O		1	
Nash, C.	SM2	65	4	
Nash, E.T.	AC1	343370	4	
Nash, H.A.	LAC	366149	6	
Nash, W.J.	AC1	505699	6	
Navin, W.P.	LAC	510474	9	
Neal, A.R.	AM2	409142	1	
Neal, B.S.	LAC	512055	9	
Neal, E.J.	F/Sgt	6999	1	
Neale, W.H.	Sgt	295435	6	
Neil, R.S.	Cpl	348043	6	
Neild, F.	AC2	83288	1	
Neilson, J.	AC2	21317	1	
Nelson, G.S.J.	LAC	364942	6	
Nelson, W.H.	F/Sgt	3153	6	
Nesbit, C.H.F.	O/O		2,3	97 Squadron
Nesbit, W.C.	AC2	162371	3	
Nesbitt, A.	LAC	362090	6	
Nettleton, A.E.	AC2	161743	(I)	27 Squadron
Nettlingham, E.C.	LAC	341577	4	
Neve, R.	LAC	365488	9	
Neville MC OBE, R.H.G.	S/L		6	60 Squadron
New, J.	CMM	1555	1	
Newbold, S.C.	ACl	347277	4	
Newbould, T.W.	Sgt	365291	9	
Newby, T.H.	AC2	155848	1	
Newcombe, L.	F/O		6	
Newham, P.S.	AC1	370779	6	
Newing, H.E.	F/Sgt	2696	4	
Newitt, J.L.	LAC	506416	8	
Newman, A.W.	AC1	344849	5	
Newman, F.	Sgt	6068	1	
Newman, H.	AC1	513021	9	
Newman, H.A.l.	LAC	362540	6	
Newman, L.l.	AC1	506471	8	
Newport, H.W.	LAC	244147	4	
Newport, S.G.	F/O		3	99 Squadron
Newry, F.J.	AC2	187406	1	
Newsom, L.H.	AC1	22733	1	
Nicholas DFC AFC, C.H.	W/C		6	
Nicholls,	AM1	78890	1	
Nicholls, R	F/Sgt	1321	4	21/24 in lieu Waz 25
Nicholls, R.F.	AC2	507001	6	
Nicholls, W.G.	F/O		4	
Nichols, W.T.H.	F/O		8	Also 36/37 and 37/39
Nicholson, D.E.O.	LAC	9097	3	
Nicholson, E.H.	AC2	330513	4	
Nicholson, G.L.	Lt		1,2,3	Special List
Nicholson, J.A.	P/O		6	
Nicholson, J.W.	LAC	340965	5	
Nieri, C.	AM1	9304	1	

Name	Rank	No.	Clasp	Comments
Nightingale, S.E.	Cpl	358659	8,9	Also 36/37
Nixon, L.G.	F/O		4	
Nixon, R.E.	P/O		1,2,3	
Noake, H.E.	AC1	512321	9	
Noble, W.C.	LAC	351933	5	
Nokes, E.H.	AC1	335587	4	
Nokes, G.A.W.	AC2	510213	6	
Nolan, F.	LAC	512995	9	
Nolder, L.J.	Sgt	347355	9	Also 36/37
Noon, L.C.	LAC	560838	9	Also 36/37
Norden, R.H.	LAC	514049	9	Also 36/37
Norris, B.J.	Sgt	365597	9	Also 36/37
Norris, C.F.	F/Lt		3	
Norris, E.	LAC	364725	9	
Norris, H.	LAC	I59113	5	
North, W.	Cpl	362100	6	
Northey, C.H.	LAC	363005	6	
Northover, W.J.	AC1	309231	3	
Northway, E.G.	F/O		6	
Norton, A.E.	Sgt	246683	6	
Norton, C.	LAC	344550	(5)	N.R.V.I.
Norton, W.E.	AM1	27081	1	
Norwood, J.	FtLt		8,9	
Nowers, F.A.	Cpl	335435	6	
Noye, J.	AC1	251072	(1)	5 Squadron
Noyes, L.B.	LAC	560839	6	
Nugent, J.	AC2	287282	1	
Nunn, C.H.N.	F/Lt		3	52 Wing
Nunneley, G.A.	Sgt	505375	6	
Oakes, T.O.	F/O		4	21/24 in lieu Waz 25
Oakley, B.A.	F/O		7	
Oakley, C.	WO2	3929	8,9	Also 36/37
Oakley, R.E.	LAC	509241	9	
O'Brien, A.	AC2	332950	(3)	60 Squadron att. 20 Squadron
O'Brien, D.	Cpl	290946	9	
O'Brien, H.H.	LAC	344957	4	
O'Brien, T.V.	AC1	342614	(5)	N.R.V.I.
O'Brien-Saint, J.T.	P/O		1,2,3	31 Squadron
O'Callaghan, P.M.	LAC	359640	6	Deceased
O'Connor, D.	LAC	507513	8	
O'Connor-Hanstock, B.J.	F/O		4	
Oddie, G.S.	P/O		1	31 Squadron
Oddie, T.	AC2	130867	(1)	27 Squadron
O'Hara, J.	LAC	364954	9	
Old, E.	AM2	18105	1	
Oldershaw, J.S.	LAC	328562	4	
Oldfield, H.A.	Cpl	361607	6	
O'Leary, A.C.	AC2	109709	(1)	20 Squadron
Oliver, G.F.	LAC	562836	9	Also 36/37
Oliver, G.R.C.	F/O		3	20 Squadron

Name	Rank	No.	Clasp	Comments
Oliver, H.F.	AM1	108197	1	
Oliver, J.	F/Lt		4	
Oliver, J.	LAC	512528	9	Also 36/37
Ommanney, M.D.	F/Lt		6	
O'Neil, W.	AC1	300491	1	
O'Neill, J.W.T.	AC2	147716	3	
Onions, J.	LAC	563359	9	
Onions, O.	AM	136898	1	
Onslow-Carleton, D.O.	F/O		3	
Onyett, K.	2/Lt		1	
Orchiston, A.J.	LAC	9897	1,2,3	
Orford, A.W.	LAC	508921	9	
Organ, C.	AC2	340789	4	
Orledge, H.H.	AC2	62005	3	
Ormerod, F.A.	LAC	359869	6	
Orriss,	AM1	164102	1	
Orton, O.	AC2	156309	1,2,3	31 Squadron att. 20 Squadron
Orton, W.H.L.	LAC	560359	9	
Osborne, A.S.M.	Cpl	243932	9	
Osborne, J.S.	LAC	9839	5	
Osborne, N.E.	LAC	298949	5	
Osborne, O P.	LAC	366374	9	
Osborne, W.H.	Sgt	9597	6	
Osbourne, G.	AM2	30758	1	
O'Sullivan, G.R.	F/Lt		9	
Otten, J.F.	LAC	247008	(5)	N.R.V.I.
Ottey, L.L.	Cpl	362548	6	
Ovenden, H.G.P.	F/O		1,2,3	
Overhill, G.	AC1	507707	6	
Owen, A.N.	AC1	343601	5	
Owen, E.O.	AC1	126761	3	52 Wing
Owen, H.J.	LAC	359616	6	
Owen, W.H.	AM2	21206	1	
Owens, F.J.	AC1	340609	3	20 Squadron
Oxenham, H.O.	AC2	225796	1	
Oxenham, W.A.	LAC	560362	9	Also 36/37 as Cpl
Oxlee, H.	AC1	370915	6	
Oxley, E.S.	P/O		1	
Paddon, B.	F/O		6	
Padwick, I.H.	LAC	363675	6	
Page, C.W.	AC2	326899	4	
Page, G.E.	LAC	341453	(5)	N.R.V.I.
Page, J.P.	Cpl	204122	4	
Page, T.J,	LAC	119522	(3)	20 Squadron
Paice, G.F.	F/Sgt	760	(3)	99 Squadron
Pain, A.C.	LAC	343141	5	
Pain, J.F.F.	F/Lt		6	
Paish, A.E.	P/O		5,9	
Palfrey, F.A.	AC1	370316	6	
Palin, V.	AC2	508333	6,9	

Name	Rank	No.	Clasp	Comments
Palliser, F.	LAC	349753	5	
Palmer, C.W.	Cpl	156397	(3)	20 Squadron
Palmer, F.R.	LAC	363192	6	
Palmer, F.W.	LAC	362598	6	
Palmer, H.F.	Sgt	349779	6	Deceased. Medal retd to Mint
Palmer, R.A.	LAC	364735	6 , 8	
Palmer, R.T.	LAC	343151	5	
Palmer, W.F.	Cpl	353088	6	
Palmer, W.J.	Sgt	242008	(1)	99 Squadron
Pannell, J.	AC1	507229	6	
Pantry, R.N.	LAC	364169	( )	
Papworth, W.	Cpl	60601	(1)	5 Squadron
Parfitt, F.E.	LAC	364741	9	
Parfitt, G.E.	Cpl	7059	1	
Pargeter, A.E.	AC1	344182	4	
Paris, W.G.R.	Cpl	343572	4	
Parish, G.	Cpl	363685	9	
Park, D.C.	AC2	343913	(3)	20 Squadron
Parker, A.L.	AC1	120255	3	
Parker, E.C.	LAC	512517	9	Also 36/37 as Cpl
Parker, M.G.	F/O		6	
Parker, R.	LAC	328214	(5)	N.R.V.I.
Parker, R.D.	Cpl	505543	8	
Parker, S.	LAC	343644	5	
Parker, S.P.	LAC	370945	6	
Parker, W.A.H.	LAC	505241	9	
Parker, W.H.	LAC	365936	8	
Parker, W.P.	SM	269	1 , 2 , 3	
Parkes, F.H.	LAC	59628	(I)	5 Squadron
Parkin, G.	LAC	561855	9	Also 36/37
Parkin, H.	Sgt	176807	1	
Parkinson, J.W.	F/O		1	
Parkinson, R.W.	Lt		1	
Parks, R.C.	Sgt	506425	6	
Parram, A.	AM2	407610	1	
Parry, B.	AC1	510232	9	
Parry, T.J.	AC2	511928	6 , 9	
Parry, T.M.	AC2	511982	6 , 9	
Parsons, A.H.	Lt		1	
Parsons, J.W.	LAC	359586	6	
Parsons, R.	AM2	409043	1	
Parsons, S.D.	LAC	362127	6	
Parsons, W.L.	Sgt	560394	9	Also 36/37 and 37/39
Part, R.F.	F/O		6	
Parton, R.E.	AC2	182739	(3), (5)	20 Squadron. N.R.V.I.
Partridge, J.	AM1	187188	1	
Partridge, J.	AC1	353291	6	
Pashley, E.A.	LAC	507663	6	
Patch, F.G.	Cpl	352974	6	
Patterson, J.	LAC	86611	(1)	5 Squadron
Patmore, S.P.A.	F/O		9	Also 36/37 and 37/39
Patrick, J.M.	Cpl	244493	5	

Name	Rank	No.	Clasp	Comments
Patston, A.J.	AC2	507251	6	
Patten, J.	LAC	359981	6	
Patterson, E.A.	Sgt	401196	(3)	20 Squadron
Pattison, H.	AC1	513528	9	Also 36/37 as LAC
Pattison, H.A.L.	F/O		1,2,3	
Patton, L.J.	AC1	251097	3,(5)	N.R.V.I
Pavitt, T.E.	F/Sgt	314995	4	
Payne, C.H.	LAC	363676	6	
Payne, C.S.	LAC	362644	6	
Payne, E.A.	LAC	363475	6	
Payne, F.L.	LAC	561849	8,9	Also 36/37
Payne, J.	Sgt	100955	6	
Payne, R.A.	LAC	510828	9	Also 36/37 as Cpl
Payne, R.N.	LAC	364418	6	
Payne, S.S.	LAC	330821	4	
Paynter, N.S.	F/O		(3)	31 Squadron
Peach, S.C.G.	LAC	505430	6	
Peach, W.	AC1	347105	4	
Peacock, H.J.	Sgt	200287	6	
Peake, G.W.	LAC	512456	9	
Pearce, A.G.	LAC	357267	6	
Pearce, A.L.	F/O		4	
Pearce, C.F.	F/O		9	Also 36/37 as F/Lt and W/C
Pearce, E.S.	LAC	355818	6	
Pearce, F.E.	Cpl	341771	5	
Pearce, F.H.	Sgt	206462	3,4	99 Squadron
Pearce, W.W.	LAC	365012	6	
Pearson, A.C.	F/O		6	
Peart, L.W.	LAC	364925	4	
Peary, E.A.E.	AC1	326856	4	
Pedley, A.	AC2	107362	1,2,3	
Peel, P.W.	Sgt	3602	1	20 Squadron
Peggs, R.	LAC	363189	6	
Peill, W.A.	AM3	155729	1	
Pelley, W.J.	LAC	355498	6	
Pelly, C.B.R.	F/O		5	
Penney, D.W.E.	LAC	362595	6	
Penniall, G.E.	Cpl	364159	9	Also 36/37
Pennicott, R.G.	SM2	224457	4	
Pennington, J.S.	LAC	365560	9	
Pennington, R.W.	LAC	362551	6	
Penny, R.W.	LAC	508331	8,9	Also 37/39 as Cpl
Penwarden, G.W.F.	LAC	359546	6	
People, S.T.E.	F/Sgt	1431	(3)	20 Squadron
Perkin, C.F.	Sgt	230436	1,3	
Perkins, A.	LAC	366154	9	
Perkins, H.	LAC	365924	6	
Perkins, W.	AC2	135097	1	Aden Flight
Perry, F.W.G.	AC1	513768	9	Also 36/37 as LAC
Perry, G.A.	AC2	251118	3	
Peskett, W.H.	Cpl	343529	4	
Petch, W.G.	LAC	565388	9	Also 36/37. 60 Squadron

Name	Rank	No.	Clasp	Comments
Peters, S.J.	LAC	365915	6	
Peters, T.A.	Cpl	336641	6	
Petersen, A.E.	Cpl	330303	6	
Petrie, W.	LAC	15530	(1)	5 Squadron
Petrie, W.G.	Cpl	119699	6	
Pett, G.A.G.	Cpl	364712	6	
Pettener, H.	AC1	511563	9	
Pettifer, A.E.C.	LAC	506929	6	
Pettifer, W.C.	AC1	33150	(3)	31 Squadron att. 20 Squadron
Pettit, T.H.	LAC	333868	(3)	52 Wing
Phelan, J.	LAC	340801	3,4	52 Wing
Phillips, E.G.	LAC	560872	9	
Phillips, E.W.	Cpl	25485	1	
Phillips, F.J.	Capt		1	
Phillips, M.A.	AC1	327158	4	
Phillips, N.V.	LAC	561339	8,9	Also 36/37
Phillips, R.H.	LAC	366442	8	
Phillips, R.L.	F/O		6	
Phillips, W.A.S.	Cpl	514384	9	Also 36/37
Philo, P.J.	AM2	29673	1	
Pickard, J.L.	LAC	505332	6	
Pickard, W.J.	F/O		6	
Pickering, W.H.	Sgt	211382	2,3	20 Squadron
Picot, E.W.J.	AC1	364162	6	
Piercy, R.	Cpl	363703	6	
Pike, T.A.	Cpl	135529	9	Also 36/37
Pilcher-Clayton, J. K.	Cpl	562588	9	Also 36/37
Pilgrim, F. B .	A/Sgt	330377	9	Also 36/37
Pilley, C.	AM	147111	1	
Pinchen, H.	Sgt	133214	6,8	
Pink, R.C.M.	W/C		5	
Pinnell, L.G.	F/O		(5)	N.R.V.I.
Pipe, E.	LAC	327750	4	
Pipe, J.	F/O		1	31 Squadron Deceased
Piper, E.M.	LAC	328172	5	
Pippett, T.F.	AC1	506933	6	
Pitman, N.C.E.	LAC	362123	6	
Pitt, C.F.	Cpl	244938	9	
Pitters, R.P.W.	LAC	365587	9	
Pittuck, W.E.S.	Cpl	349318	(5)	N.R.V.I.
Plant, W.R.	AC1	404239	1	
Platford, A.E.	F/O		4	21/24 in lieu Waz 25
Platt, J.	P/O		1,2,3	
Platt, T.H.	A/Sgt	6392	6	
Platten, A.G.	F/Sgt	21460	1	
Playford, E.R.B.	F/Lt		4	
Pleasants, R.	LAC	507720	6,8	
Plenderleith, R.L.	AC1	10701	3	
Plenderleith, W.N.	F/Lt		7	
Plumbridge, H.G.	Sgt	110871	3	
Plummer, W.	AC1	513753	9	
Plumridge, H.G.C.	2/Lt		1	

Name	Rank	No.	Clasp	Comments
Plunkett, W.J.	Cpl	361730	6	
Pollard, E.W.	AC1	144542	(1)	5 Squadron
Pond, W.J.	Cpl	349798	5	
Pond, W.T.	AM2	48247	1	
Pook, E.G.	LAC	102135	(1)	5 Squadron
Poole, W.H.	F/Lt		3	
Pope, A.E.	Cpl	362583	6	
Pope, J.C.	F/O		9	
Pope, S.C.	AC1	370547	6	
Porter, E.F.	F/O		8,9	
Porter, H.P.	AC1	248053	4	
Porter, J.C.	AC2	510518	6	
Porter, W.J.	2/Lt		1	
Porter, W.R.	Cpl	340283	6	
Pothecary, W.	AM2	136894	1	
Potten, W.H.	Sgt	362602	6	
Potter, E.J.	AC1	66573	(1)	20 Squadron -
Potter, S.	AC1	247092	1	
Potts, A.E.	AC2	91352	( )	5 Squadron
Pound, S.G.S.	AC1	349510	5	See 'R.Soder'.
Poutney, W.	AC	359764	6	
Powe, S.	Cpl	344080	( )	
Powell, A.	LAC	63788	2,3	
Powell, A.A.E.	LAC	362115	6	
Powell, E.	AM2	103994	1	
Powell, E.	Cpl	108969	1	
Powell, F.	LAC	365015	6	
Powell, F.J.	F/Lt		3,4	28 Squadron
Powell, H.S.	S/L		4	
Powell, R.R.	Sgt	352007	9	Also 36/37
Powell, T.S.	LAC	356974	6	
Power, A.H.	F/O		2,3	97 Squadron
Power, A.G.	LAC	561512	( )	Also 36/37 as Cpl
Power, T.C.	Sgt	342506	4	
Powles, A.H.	Cpl	362129	6	
Powley, A.G.	LAC	511548	9	
Pratley, T.C.	LAC	366153	9	
Pratt, C.V.J.	F/O		6	
Pratt, F.	LAC	364940	6	
Prendergast, A.R.	F/O		5	
Prendergast, F.M.	AC1	509263	6	
Prescott, W.A.	AC1	342051	4	
Preston, J.	Cpl	349323	8,9	
Preston, W.C.	LAC	326182	5	
Pretty, H.J.	Sgt	4716	1	
Prevost, F.G.	AC1	165765	(1)	5 Squadron
Prewitt, E.F.	Sgt	404	(3)	20 Squadron
Price, E.J.	AC1	249313	(3)	
Price, G.	Sgt	178432	(3)	
Price, J G,	Sgt	340475	9	
Price, L.	AM2	13574	1	
Price, L.G.	LAC	511160	9	

Name	Rank	No.	Clasp	Comments
Price, S.E.	LAC	335024	(5)	N.R.V.I.
Price, S.J.	Cpl	334013	6	
Prickett, W.W.	AC1	365117	6	
Prideaux, W.G.E.	LAC	330254	(5)	N.R.V.I.
Priest, A.E.	Sgt	363473	6	
Priest, C.T.	AM2	29197	1	
Prince, J.W.	Cpl	365977	9	Also 36/37
Pringle, W.	LAC	351857	4	
Prior, A.S.	F/Sgt	315048	4	
Pritchard, H.	Sgt	30703	1	
Pritchard, H.G.	Cpl	390128	9	Also 36/37
Pritchard, O.F.	Sgt	236478	4	
Proctor, A.E.	LAC	327612	4	
Proffit, J.G.	LAC	365113	6	
Proud, H.	LAC	343385	4	21/24 in lieu Waz 25
Pryce, E.T.	AC1	513263	9	
Pryor, A.D.	F/Lt		2,3	99 Squadron
Pugh, W.E.	Sgt	17413	1	
Pugh, W.J.	LAC	362108	6	
Pughe, R.	P/O		1	Deceased
Puleston, C.	LAC	357451	6	
Pulling, W.H.C.	LAC	327302	4	
Punchard, H.A.	AC2	397474	4	
Purdin, W.E.	F/O		1	
Purdy, J.	LAC	505171	( )	
Purdy, L.	AC1	327520	4	
Purssell, T.R.L.	LAC	330157	4	
Putt, W.C.	Cpl	363791	6	A/Sgt
Pyke, A.	F/O		8,9	
Pyne DFC, R.	F/O		4	21/24 in lieu Waz 25. 31 Sqdn
Quartermaine, R.F.	Cpl	362137	6	
Quayle, E.S.	LAC	331337	(3)	52 Wing
Quennell, H.J.	LAC	508530	9	Also 36/37 as Sgt
Quennell, W.C.	LAC	510950	8,9	
Quested, J.B.	Major		1	48 Squadron
Quick, C.J.M.	AC2	312527	3	
Quick, F.G.	AC2	337081	3	20 Squadron
Quick, L.	LAC	363235	6	
Quinton, F.	AC2	333943	3,(5)	N.R.V.I.
Quirk, A.J.	LAC	348647	6	
Quirk, G.L.	LAC	363467	6	
Raby, B.	Sgt	247530	3	28 Squadron
Radcliffe, J.C.	LAC	327485	4	21/24 in lieu Waz 25
Raddon, A.J.	LAC	512822	9	Also 36/37 as Cpl
Radford, J.	LAC	240374	4	
Railton, J.J.R.	AC1	505704	6	
Rainsbury, S.T.	Sgt	313873	6	
Ralph, R.F.	P/O		1,2,3	

Name	Rank	No.	Clasp	Comments
Ralphs, W.J.	F/Sgt	326013	9	Also 36/37 as W/O
Ramm, H,J,	Cpl	341034	5	
Rampton, P.T.	AM2	404299	1	
Ramsay, A.C.	AM	406249	1	
Ramden, E.	WO1	1216	1,3	Incorrectly engraved.
Randall, G.E.	F/O		1	20 Squadron. Medal engraved 'G. Randall'
Rankin, R.C.	LAC	509338	6	
Rankin, R.M.	F/O		1	
Rankine, S.M.	LAC	248647	5	
Ranson, H.G.	AC1	510190	8,9	Also 36/37 as LAC
Raspison, L.G.	Cpl	335789	9	Also 36/37 as Sgt
Ratcliffe, C.H.	F/O		4	
Ravenscroft, H.E.	LAC	358245	6	
Rawlinson, G.H.	F/O		5	
Rayman, V.	AM1	22901	1	
Rayner, G.J.	F/O		4	28 Squadron
Raynor, W.A.	Sgt	143072	1	
Read, G.H.	LAC	590107	9	
Read, L.F.	LAC		1	
Read, W.J.	AC2	515299	9	Also 36/37 as AC1
Reader, J.W.G.	LAC	358806	6	
Readhead, A.W.	AC1	364238	6	
Readhead, W.F.	AC1	363236	6	
Readman, A.	Cpl	345259	4	
Reaney, M.A.	W/O	94659	(8)	N.R.V.I.
Reardon, W.J.	AC2	513844	8,9	Also 36/37 as Cpl
Reay-Jones, R.	F/Lt		9	
Redgewell, G.E.	LAC	364113	6	
Redhead, J.J.C.	AC1	352712	5	
Redman, A.	AM2	37358	1	
Reed, A.J.	AC1	342699	5	
Reed, H.S.	AC1	370136	6	
Reed, J.R.	LAC	508320	9	
Rees, F.A.	LAC	362143	6	
Rees, H.G.P.	F/Lt		3	
Rees, R.D.J.	LAC	363239	6	Deceased
Rees, W.C.	AC1	560390	6	
Reeve, S.L.	Cpl	337185	5	
Reeves, A.J.	LAC	363011	6	
Reeves, H.S.C.	LAC	356275	6	
Reeves, W.E.	AC2	148727	1	
Reid, A.C.	LAC	338019	4	
Reid, H.R.McL.	F/Lt		9	Also 36/37 as S/L
Renfrew, P.	AM1	408770	1	
Rennie, W.	AC2	332724	(3)	99 Squadron
Rennolls, J.G.	LAC	87945	(3)	20 Squadron
Renshaw, C.M.B.	F/O		9	Also 36/37 - N.R.V.I.
Rew, H.E.	F/O		4	Lt R. Berks Regiment
Reynolds, A.E.	O/O		1,3	Medal O.C.
Reynolds, J.V.J.	LAC	507899	6,9	
Reynolds, L.H.	LAC	510983	8,9	Also 36/37

Name	Rank	No.	Clasp	Comments
Reynolds, R.J.	LAC	561340	9	
Reynolds, S.	AC2	141956	1	
Reys, M.E.	AC1	364204	6	
Rich, H.T.	LAC	366372	9	
Rich, R.T.	O/O		3	
Rich, S.C.R.	Cpl	343633	4	21/24 in lieu Waz 25
Richards,	AM1	12918	1	
Richards, F.J.	AM2	22813	1	
Richards, H.G.	F/O		6	
Richards, J.	LAC	341987	4	
Richards, W.J.	F/O		4	
Richards, W.K.	Sgt	348789	6	
Richardson, A. C.	F/O		6	
Richardson, C. J.	AC1	351485	4	
Richardson, E.H.	F/O		(3)	20 Squadron
Richardson, G.T.	F/O		2	20 Squadron
Richardson, H.	AC1	370010	6	
Richardson, I.B.	LAC	358549	( )	
Richardson, J.E.	AC1	326590	5	
Richardson, L.J	Cpl	346543	6	
Richardson, S.E.H.	LAC	356580	6	
Richardson, T.	AC1	344711	4	21/24 in lieu Waz 25
Riches, D.A.	LAC	156588	5	
Richmond, G.N.	P/O		1,2,3	
Rickard, A.G.	Sgt	352193	9	Also 36/37 as F/Sgt
Rickards, P.W.	AM1	408963	1	
Rickerby, B.R.	AC2	326817	(3),5	20 Squadron
Ricketts, W.A.	AC2	126229	3	
Ricketts, W.J.	AC2	561108	9	
Ridealgh, C.	Sgt	328246	4	
Ridgley, H.J.	AC1	252554	1,3	
Ridley, G.	AC1	55420	1	
Ridley, L.H.	F/O		4	
Rigg, R.O.	F/O		(5)	N.R.V.I.
Riggs, T.H.R.	F/O		4	Now Lt in Lincs. Regiment
Riley, H.	AC1	509160	8,9	
Riley, I.C.	AC2	303943	1,3	
Riley, J.H.	A/Cpl	12959	1	
Riley, P.	AC1	506781	6	
Ring, R.F.	AC1	506713	6	Deceased
Rippon, T.S.	W/C		1, 3	
Riversdale- Elliott, K, H.	S/L		9	Also 36/37 as W/C
Road, A.J.	AM1	19278	1	
Robb, J.M.	LAC	327656	5	
Robbins, S. G .	LAC	341974	4	
Roberts, A.F.	Cpl	358531	6	
Roberts, C.J.	AM2	405878	1	
Roberts, C.V.	LAC	507476	6	
Roberts, C.Y.	F/Lt		4	
Roberts, E.	LAC	564395	9	Also 36/37
Roberts, F.E.C.	LAC	364988	9	Also 36/37

Name	Rank	No.	Clasp	Comments
Roberts, G.N.	F/O		6	
Roberts, G.W.	AC1	350550	4	
Roberts, J.E.	2/Lt		1	
Roberts, J,H.	LAC	560382	9	Also 36/37 as Cpl
Roberts, M.J.	Cpl	157700	1,3	20 Squadron
Roberts, R.A.P.	F/Lt		6	
Roberts, R. C.	LAC	506400	6	
Roberts, R.T.	AC1	356439	6	
Roberts, T.G.	LAC	510090	8	
Roberts, T.R.	LAC	359121	6	
Roberts, V.H.	LAC	505199	9	
Robertson, E.A.J.	LAC	513636	9	Also 36/37 as Cpl
Robertson, R.G.	Sgt	366169	9	Also 36/37
Robertson, R.S.	F/Sgt	242557	9	
Robins, A.S.Q	F/O		9	Also 36/37 as F/Lt
Robins, P.W.	LAC	327686	5	
Robins, W.R.	Cpl	86779	4	Also took part in Waz 25 Opns
Robinson, A.	AC1	85638	(3)	99 Squadron
Robinson, A.J.	AM1	3833	1	
Robinson, A. R.	LAC	363470	6	
Robinson, E.W.	AC1	245485	(5)	N.R.V.I.
Robinson, F.A.	Sgt	334098	6	
Robinson, F.G.A.	F/O		2,3,6	
Robinson, H.P.	AC2	233574	3	
Robinson, H.T.	Sgt	362662	6	
Robinson, J.S.	Sgt	562270	9	Also 36/37
Robinson, L.	AC1	329240	2,3	Alias Gonzalez
Robinson, T.	LAC	507239	9	Also 36/37 and 37/39 as AC1
Robinson, W.A.J.	AC2	342226	5	
Roblou, T.	AC1	363510	6	
Roby, S.	AC1	334826	4	
Rodber, R.H.	F/Sgt	330832	9	Also 36/37
Rodgers, G.	AC1	513016	9	Also 36/37 as LAC
Rodgers, J.F.	AC1	512241	9	Also 36/37 as LAC
Roe, C.H.	LAC	336597	6	
Roe, M.	LAC	370981	6	
Rogers, A.C.T.	AC2	508489	6	
Rogers, E.P.	LAC	362764	6	
Rogers, E.T.	AC2	139366	3	Now 5566269 Pte Wilts Regiment
Rogers, F.	AC1	332621	3	
Rogers, F.C.W.	LAC	365336	6	
Rogers, G.W.	Sgt	244733	6	
Rogers, J.	LAC	366408	9	
Rogers, J.T.	LAC	366168	9	
Rogers, W.C.G.	AC1	514194	9	Also 36/37 as Cpl
Rogerson, G.	Lt		1,2,3	
Rolfe, C.V.	Cpl	78592	1	5 Squadron
Rolfe, G.	AC1	510029	8,9	
Rolfe, T.U.	F/O		9	
Rollins, F.E.P.	LAC	512276	9	Also 36/37 as Cpl
Rollinson, W.	AC1	370827	6	
Ronan, J.H.	AC2	239171	I	

Name	Rank	No.	Clasp	Comments
Rooke, D.M.	Lt		1	
Rooney, J.	LAC	513506	9	
Rootes, J.T.	AC1	358923	6	
Rosam, F.R.	LAC	365320	9	
Rose, C.	AC1	516163	9	
Rose, J.	AM2	187863	1	
Rose, J.	AC1	340982	(5)	N.R.V.I.
Rose, S.A.W.	Cpl	243677	4	21/24 in lieu Waz 25
Rose, T.	F/O		2,3	99 Squadron
Rose, T.J.	Cpl	355610	6	
Rose, W.R.	LAC	365932	6,8	
Rosenheim, A.G.	AC2	94060	1,2,3	
Ross, A.	LAC	327914	4	
Ross, A.R.	Sgt	366421	9	Also 36/37
Ross, J.	AC1	83445	4	Not eligible for Waz 19/21
Ross, J.A.G.	LAC	365791	9	
Ross, M.St.J.	P/O		1	Medal O.C.
Rotheram, E.	F/O		6	
Rothery, C.	LAC	348212	4	
Round, J.Y.	F/Sgt	341685	6	
Row, W.T.H.	LAM	20049	1	
Rowbotham, E.	Sgt	354659	9	
Rowe, H.	LAC	561621	9	Also 36/37 as Cpl
Rowe, H.J.	Sgt	334158	6	
Rowe, H.W.	Cpl	328195	6	
Rowe, L.G.	AC1	505843	6	
Rowe, O.R.	F/Sgt	230	1,3,4	20 Squadron
Rowe, S.W.	AM3	149724	1,2,3	
Rowe, T.A.	LAC	508932	8,9	
Rowe, T.W.	Cpl	12915	1	
Rowell, R.F.	F/Sgt	347924	9	Also 36/37. Now F/O
Rowell, W.A.E.	LAC	342713	4	
Rowlett, C.F.	LAC	366165	9	
Rowlinson, A.	LAC	560891	9	
Roxburgh, W.A.L.	AC1	352045	4	
Roy, C.	AC1	342394	5	
Royce, J.	LAC	366407	9	
Rubio, E.	Cpl	351936	6	
Ruddock, G.E.	Sgt	326373	9	
Rugg, C.W.	F/O		5	
Ruggles, R.J.	AC2	163596	1,3	20 Squadron
Rumbelow, E.	LAC	342882	(4)	N.R.V.I. Also in Waz 25 Opns
Ruse, W.L.	AC1	505631	6	
Russell, B.H.C.	P/O		1,2,3,4,6	48 Squadron
Russell, H.E.W.	LAC	366177	9	
Russell, J.C.	S/L		1,3	20 Squadron
Russell, J.H.	LAC	562609	9	
Russell-Stracey, E.G.H.	F/Lt		6	
Ruston, A.G.	Cpl	353959	6	
Rutherford, A.D	Sgt	7392	3,(5)	20 Squadron. N.R.V.I.
Rutherford, J.	LAC	507328	6,8,9	
Ryan, J.	LAC	305653	4	

Name	Rank	No.	Clasp	Comments
Ryan, S.E.L.	Cpl	359997	6	
Ryder, G.	Cpl	560893	9	Also 36/37 as A/Sgt
Ryley, D.W.R.	F/O		6	
Sadler, F.W.	AC2	83432	1	5 Squadron
Sadler, L.R.	LAC	347635	5	
Saffery, J.H.C.	LAC	363748	6	
Saggers, A.C.	AM1	21233	1	
Sainsbury, F.W.	LAC	363711	6	
Saker, H.J.	F/O		2,3	
Salaman, G.H.	Capt		1	
Salmon, F.T.	LAC	363326	6,8	
Salmon, H.S.	AC1	560898	6,8,9	
Salter, H.B.	Cpl	365590	9	Also 36/37
Salter, W.E.	Cpl	590051	9	
Sammels, A.J.H.	LAC	364254	6	
Samson, R.A.	LAC	362646	6	
Sancto, A.G.	Sgt	364446	8	
Sanders, G.F.	Sgt	348486	9	
Sanderson, A.C.	F/Lt		5	
Sanderson, E.G.	LAC	366463	9	
Sanderson, F.S.	AC2	409041	1,3	
Sanderson, W.E.	AC2	511455	6,8,9	
Sandison, J.R.O.	LAC	363035	6	
Sandow, J.F.	F/Lt		9	Also 36/37
Sandy, A.R.	AC1	326505	4	21/24 in lieu Waz 25
Sandys, H.E.	AC1	510988	8,9	
Sanger, W.	AM2	7162	1	
Sansbury, R.H.	AC2	506839	6	
Sarjeant, H.W.	Cpl	252781	4	21/24 in lieu Waz 25
Saunders, C.F.	LAC	362177	6	A/Cpl
Saunders, C.R.	Cpl	348053	5	
Saunders, E.	AM2	13433	1	
Saunders, E.M.	AM2	136896	1	
Saunders, R.J.	LAC	590292	9	Also 36/37 as Cpl
Saunders, W.T.	LAC	509216	9	
Savage, G.	LAC	342765	4	
Savery, R.C.	F/Lt		5	
Savill, G.	F/Sgt	313892	4	
Saward, N.C.	F/O		(3)	28 Squadron
Sawyer, H.	AC1	144830	4	21/24 in lieu Waz 25
Sawyer, T.J.	LAC	13602	1	20 Squadron
Saxby, J.H.	AC1	359749	6	
Saxby, W.H.	Cpl	333824	(3)	20 Squadron
Sayer, J.E.	AC1	515434	9	Also 36/37 as LAC
Sayer, M M	LAC	510595	9	
Sayle, E.	AC2	66156	1	
Sayner, W.	Sgt	4902	1	
Scarfe, W.	AM1	16053	1,2	
Scarrott, G.	F/O		2,3	20 Squadron
Schasche, D.H.	LAC	328603	(5)	N.R.Y.I.

Name	Rank	No.	Clasp	Comments
Schofield, F.G.	Cpl	344181	6	
Scholefield, R.	Sgt	251939	6	
Scholey, W.H.	AC2	348168	4	21/24 in lieu Waz 25
Schoolar, W.O.J.	AC1	511269	9	
Schroder, H.G.	LAC	35S397	6	
Scivier, D.W.	Sgt	560689	9	
Sclater, G.C.	F/O		4	
Scoble, N.W.S.	LAC	364274	6	
Scott, A.B.	LAC	334136	5,9	
Scott, A.G.	LAC	356602	8	
Scott, C.E	AM2	78888	1	
Scott, G.A.	LAC	330400	5	
Scott, G N.	AC2	122823	2,3	
Scott, P.G.	F/Lt		4	
Scott, S.L.	Cpl	335583	4	
Scraton, J.H.	Sgt	4888	1	
Scriven, W L	AC1	89200	(3)	97 Squadron
Scrivin, P.J.	LAC	370601	6	
Seaman, H.T.	Cpl	25694	4	21/24 in lieu Waz 25
Searle, S.G.	Cpl	560909	9	
Seath, F.A.	AC2	341150	4	
Seddon, H.	LAC	90618	1	
Seeley, W.H.	LAC	508505	9	
Segrove, J.E.	AC2	346579	4	
Self, C.R.	AC1	344367	4	
Sellek, E.S.	Cpl	159973	5	A/Sgt
Selman, H.A.	Sgt	12889	1	
Semley, F.	AC2	334617	(3)	20 Squadron
Senier, H.	LAC	345600	4	
Senior, A.	LAC	345419	6	Deceased
Senior, R.C.	Cpl	11702	1	
Seward, W.J.	F/Lt		1	
Sexton, D.R.	Sgt	345237	9	
Seymour, A.E.	LAC	370245	6	
Shackell, E.A.	LAC	513082	9	
Shaddick, E.J.	LAC	362635	6	
Shadwell, C.	Cpl	6120	1	
Shannon, J.R.	AM2	13127	1	
Shapcott, M.S.	F/O		6	
Sharp, A.C.H.	F/Lt		8	
Sharp, A.E.	AC2	337186	(3)	20 Squadron
Sharp, J.	AC1	355741	6	
Sharpe, O.E.	F/O		2,3	99 Squadron
Sharpe, S.F.	LAC	370190	6	
Sharwood-Smith, B.E.	Lt		1	
Shaw, E.	AC2	340843	4	21/24 in lieu Waz 25
Shaw, G.H.	S/L		9	Also 36/37. Deceased
Shaw, L.R.	P/O		5	
Shaw, R.T.	LAC	515635	9	
Shaw, S.A.J.	Sgt	358196	9	
Shaw, T.J.	F/O		4	
Shaw, W.	LAC	564426	9	Also 36/37 as Cpl

Name	Rank	No.	Clasp	Comments
Shears, F.	Cpl	8011	1	
Shears, R.H.	F/Lt		1	31 Squadron
Sheath, L.W.	LAC	363253	6	
Sheffield, W.H.	AM3	22127	1	
Sheldrick, F.	Sgt	23323	1	20 Squadron
Shelley, G.N.	AC1	342628	4	
Shepherd, A.T.H.	AC1	345767	4	
Shepherd, W.	AM2	24083	1	
Sherman, G.B.	LAC	511846	9	Also 36/37 as Cpl
Sherren, P.C.	S/L		4	
Sherwin, C.H.	F/Lt		1	31 Squadron
Sherwin, C.J.	AM2	113035	1	
Sherwood, W.	Cpl	6691	1	
Sheward, J.A.	AC1	363485	6	
Shiers, W.H.	Sgt	27	1	
Shone, L.	AM1	82005	1	
Short, J.P.	LAC	511061	9	
Short, M.E.H.	AC1	330170	5	
Short, R.W.	Sgt	235411	3	20 Squadron
Shoubridge, M.W.	LAC	358790	6	
Shufflebotham, J	AC2	302577	1	
Sidlow, F.G.	AC1	514047	9	Also 36/37 and 37/39. Now F/Lt
Sigsworth, J.F.	Cpl	180857	5	
Silversides, C.A.	Cpl	346956	9	
Silvester, F.J.	Sgt	244940	9	Also 36/37 as F/Sgt
Simister, G.H.	F/Lt		1	
Simkins, J.	LAC	24571	1	
Simmonds, A.D.	AC2	344305	5	
Simmonds, H.W.	AC1	233950	(3)	
Simmonds, J.P.H.	LAC	363322	6	
Simmons, A.	LAC	370602	6	
Simmons, J.	AC1	157631	(3)	99 Squadron
Simms, H.J.	LAC	365951	6	
Simonds, W.S.P.	F/O		9	Also 36/37
Simons, F.H.	AM2	11614	1	
Simpkins, T.	AC1	514291	9	Also 36/37
Simpson, A.E.	LAC	506675	9	
Simpson, B.H.	LAC	507003	6	
Simpson, C.A.	LAC	561359	9	Also 36/37 as Cpl
Simpson, E.E.	AC1	516228	9	
Simpson, F.H.	LAC	408831	1	Attached I/7 Batt R.F.A.
Simpson, J.H.	AC1	506985	6,8	
Simpson, R.	LAC	512466	9	
Simpson, W.S.	T/Sgt	55150	2,3	
Sims, F.	AC2	293470	1	
Sims, J,H.	AC1	359605	6	
Sinclair, A.D.	Lt		3	Duplicate medal issued
Sinclair, E.A.	AC1	365342	6	
Sinclair, L.F.	F/O		6	28 Squadron (1927)
Singer, W.	AC1	342791	4	
Singleton, W.J.	Cpl	361636	6	
Siviter, T.V.	AM3	159339	1	

Name	Rank	No.	Clasp	Comments
Skelly, G.I.	AC2	294877	3	
Skene, J.B.	AC2	342999	5	
Skene, W.E.	LAC	560910	9	Also 36/37 as Cpl
Skidmore, A.C.R.	Cpl	19552	9	Also 36/37 as A/Sgt
Skiller, D.R.	LAC	590351	9	
Skillings, E.J.	LAC	342l14	4	21/24 in lieu Waz 25
Skilton, A.F.	AM2	11609	1	
Skingsley, E.R.	Sgt	334337	6,8	
Skinner, A.	LAC	506750	8	
Skinner, B.J.	F/Sgt	337212	6	
Skinner, S.	AC1	348471	5	
Skivington, J.	AC2	342210	4	
Skoulding, F.A.	F/O		4	
Slack, J.T.	Cpl	342158	6	
Slade, S.C.	LAC	361614	6	
Slark, H.E.	AC1	506119	6,8	
Slater, A.N.	AC1	560902	6	
Slater, N.	LAC	358501	6	
Slatter, C.J.	LAC	590119	9	
Slatter, H.	LAC	563232	9	
Sloane, J.F.	LAC	511972	9	
Slorance, A.	LAC	560414	9	
Slynn, R.	AC1	506453	6	
Small, C.W.	AM1	5218	1	
Small, G.R.	Cpl	334124	5	A/Sgt
Smart, C.F.	LAC	364762	6,9	
Smith, A.	AM3	37250	1	
Smith, A.	Sgt	26785	4	
Smith, A.A.	AC1	157751	4	
Smith, A.E.	F/Sgt	634	1	
Smith, A.E.	LAC	328152	(5)	N.R.V.I.
Smith, A.E.	LAC	361640	6	
Smith, A.H.J.	Cpl	357129	8	
Smith, A.W.	AC1	358713	6	
Smith, A.Y.	LAC	347410	4	
Smith, C.	LAC	358773	6	Deceased
Smith, C.F.	Cpl	342046	4,6	
Smith, C.W.	LAC	365316	6	
Smith, C.W.	AC1	335159	5	
Smith, E.J.	LAC	342746	4	
Smith, E.P.	Cpl	356253	6	
Smith, E.T.	F/O		9	
Smith, E.W.	LAC	366413	9	
Smith, F.	Sgt	313906	4	
Smith, F.	LAC	510958	9	
Smith, F.A.	LAC	249012	4	
Smith, F.A.R.	F/O		4	21/24 in lieu Waz 25
Smith, F.G.	LAC	510663	9	28 Squadron (1932)
Smith, F.H.	Cpl	352137	9	
Smith, F.H.	LAC	352819	5	
Smith, F.W.	AM2	406633	1	
Smith, G.A.	Sgt	362671	6	Not entitled Mohmand 1933

Name	Rank	No.	Clasp	Comments
Smith, H.A.	AM2	406227	1	
Smith, H.C.	Sgt	364783	9	
Smith, H.E.	AC1	326531	4	
Smith, H.F.	AC1	249401	(5)	N.R.V.I.
Smith, H.W.	Cpl	340929	6	
Smith, J.	LAC	352877	6	
Smith, J.	LAC	26328	1	114 Squadron
Smith, J.K.	F/O		4	
Smith, J.W.	LAC	506533	6	
Smith, J.W.	AC1	356050	6	
Smith, L.H.	AC1	345413	4	
Smith, L.R.	LAC	359754	9	
Smith, N.I.	F/Lt		6	
Smith, P.G.	AC2	134393	1	
Smith, P.H.	AC1	363699	6	
Smith, R.J.	Lt		1	
Smith, S.C.	AC2	344754	4	21/24 in lieu Waz 25
Smith, S.L.	AC1	511930	8	
Smith, S.W.E.	F/Sgt	349966	9	Also 36/37
Smith, T.E.	AM1	28590	1	
Smith, W.	Cpl	355849	6	
Smith, W.	Sgt	20948	1 , (3)	
Smith, W.T.	LAC	341027	4	A/Cpl
Smith, W.T.J.	LAC	561631	9	
Smythe, A.	AC1	291688	6	
Smythe, C.R.	F/O		(3)	20 Squadron
Smythies, B.E.	S/L		(3)	
Snary, H.	AC	92891	1	114 Squadron
Snell, H.A.	LAC	364260	6	
Snell, H.C.	LAC	355036	6	
Snell, W.B.	AC2	332557	(3)	20 Squadron
Soars, A.E.	AC2	507641	6	
Soden, F.O.	F/Lt		(3)	20 Squadron
Soder, R.	AC2	349573	5	Returned unclaimed. Used for 349510 AC1 S.G.S. Pound
Somervell, W.E.	F/Lt		4	
Soper, G.F,	LAC	365616	6	
Soper, H.J.	LAC	365615	6	
Southcott, G.E.	A/Cpl	67503	(1)	20 Squadron
Southgate, R.W.	A/Sgt	33S608	9	
Southgate, W.J.	SM1	753	2	
Spacey, W.J.	LAC	16190	1	
Spackman, A.	AC1	326803	5	
Spackman, J.H.S.	LAC	358521	6	
Sparks, L.A.	AC1	239481	(1) , 6	
Sparks, W.G.	Sgt	157549	4	
Sparrow, H.	AM2	148735	1	
Spears, B.	AC1	211948	I	
Speight, A.	AM1	21832	1	
Spence, E.H.D.	F/O		5	
Spence, J.S.	AC1	327040	4	
Spencer, D.G.H.	P/O		9	Also 36/37 and 37/39. 20 Sqdn

Name	Rank	No.	Clasp	Comments
Spencer, G.	LAC	349276	4	21/24 in lieu Waz 25
Spencer, G.R.C.	F/O		5	
Spencer, S.	Sgt	362148	6	
Spicer, O.C.	AMI	187899	1	
Spiller, W.	AM2	9448	1	
Spittlehouse, J.	LAC	561374	9	Also 36/37
Spokes, F.J.	ACI	343761	(3)	20 Squadron
Spooner, A.C.	LAC	505453	6	
Spooner, F.	LAC	326747	5	
Spooner, P,	ACI	505702	6	
Spoors, E.L.	LAC	560912	8 , 9	
Spreckley, H.D.	F/O		6	
Sprague, P.T.	AC2	344864	4	
Spreadborough, F.T.	Cpl	3498	4	21/24 in lieu Waz 25
Springate, A.	F/Sgt	410	3 , (5)	
Springett, G.J.	LAC	561387	9	Also 36/37
Springham, H.W.C.	Sgt	205852	(3)	52 Wing
Spurden, F.G.	Sgt	359221	9	
Spurrier, H.	AC1	37955	1	
Squire, G,L.	LAC	363706	6	
Squirrell, W.R.	Cpl	350972	8	
Stacey, A,N.	LAC	364285	6	
Stafford, T.W.	Cpl	7413	1	
Stagg, W.A.	F/O		6	
Stallard, H.H.	AC1	337853	4	
Stammers, R.E.	AC1	515443	9	Also 36/37 and 37/39 as LAC
Stamp, E.H.	Sgt	505357	6	
Standen, H.L.	Cpl	159207	1	
Stanger, W.T.	LAC	249536	4	
Stanley, A.G.	LAC	362161	6	
Stannard, F.W.	F/O		6	
Staples, G.N.	LAC	364267	6	
Stapleton, W.	AM1	148010	1	
Stapley, J.H.	Sgt	6684	1	
Stark, W.	AM2	13727	1	
Staton, W.E.	F/O		2,3	20 Squadron
Stead, F.	Cpl	109021	(1)	5 Squadron
Stead, J.	AC2	340075	3	
Stead, P.J.	LAC	357979	6	
Steadman, F.C.	Sgt	207662	4	
Stears, A.D.S.	AC1	232646	4	21/24 in lieu Waz 25
Stedman, R. de L.	F/Lt		I	
Steel, A.	Cpl	32076	I	
Steel, A.P.	Sgt	363278	6	
Steel, H.H.	LAC	561427	9	
Steele, C.R.	F/O		3,4	20 Squadron
Steele, M.	F/Sgt	244403	6	
Steele, Q.C.	LAC	563707	9	Also 36/37 and 37/39
Steele, R.P.	AC1	S06262	6	
Steer, A.E.	LAC	370054	6	
Steers, E.A.	F/Sgt	149978	6	
Stentiford, O.C.	LAC	363323	6	

Name	Rank	No.	Clasp	Comments
Stephens, E.J.	SM2	338011	6	
Stephens, W.C.	LAC	343027	(5)	N.R.V.I.
Stephenson, A. C.	F/Sgt	27600	6	
Stephenson, G.C.H.	Cpl	327953	9	Also 36/37 as Sgt
Stephenson, G.H.W.A.	LAC	511503	9	
Stevens, C.A.	F/Lt		5	
Stevens, F.O.	AM2	28389	1	
Stevens, H.	AC1	343124	5	
Stevens, H.G.	AC2	164182	(I)	27 Squadron
Stevens, S.	AM2	12366	I	
Stevens, T.G.	Cpl	361616	9	
Steward, S.F.	Cpl	345111	6	
Stewart, A.J.	LAC	335763	6	
Stewart, C.M.	F/O		9	
Stewart, G.V.	F/O		1	31 Squadron
Stewart, I.G.	LAC	365815	6 , 9	
Stewart, J.	Cpl	560210	9	Also 36/37
Stewart, J.D.	LAC	507746	6	
Stewart, L.E.C.	LAC	364256	6	
Stewart, W.	SM1	630	I	20 Squadron
Stewart, W.	P/O		I	
Stewart, W.H.	F/Sgt	408782	1	
Stiles, R.J.	Cpl	353880	9	
Still, C.G.	LAC	560416	9	
Stillman, W.J.	LAC	513511	9	Also 36/37. Now P/O
Stillwell, H.E.	LAC	345404	4	
Stitchall, W.	AC2	508072	6	
Stobart, L.	LAC	264478	4	21/24 in lieu Waz 25
Stobe, A.	LAC	15389	1	
Stobo, A.	LAC	15837	3	
Stocks, R.B.	Sgt	364172	6	
Stodart, D.E.	S/L		1	114 Squadron. Medal O.C.
Stoddart, J.S.	E AC1	370281	6	
Stokes, G.S.	AC1	512405	9	
Stokes, L.R.	F/Lt		9	
Stone, E.H.	LAC	510994	9	
Stone, G.J.A.	LAC	63609	(3)	
Stone, W.E.	Cpl	327451	4	
Stonebank, W.	Sgt	347138	9	
Stonebridge, J.	Cpl	12713	1	
Stonier, R.	Cpl	357922	9	
Storer, H.H.R.	LAC	362168	6	
Story, R.E.	AC2	506220	6	
Strachan, G.S.	F/Lt		6	
Straight, W.H.	LAC	332911	(3)	99 Squadron
Strange, R.B.	AC1	87293	4	
Strath, F.A.A.H.	F/O		9	
Stratton, J.A.C.	F/O		6	
Street, A.V.	Lt		1	
Street, C.K.	LAC	365959	7	
Street, E.J.	Sgt	365308	9	
Street, F.G.		348651	5	

Name	Rank	No.	Clasp	Comments
Street, F.H.	AC2	508386	6, 8	
Street, W,	AC2	118545	(1)	5 Squadron
Streeter, P.R.	Cpl	356285	6	
Strevens, B.	AC2	19201	1	2O Squadron
Strevens, L.J.	LAC	370030	6	
Strong, G.E.	LAC	511078	9	
Strong, H.	Sgt	9451	1	
Stroud, R.W.A.	F/O		6	Deceased
Strudwick, E.W.	F/Sgt	313923	5	
Stuart, L.J.	Capt		1	
Stubbins, J.	AC2	507607	6	
Stubbington, H.J.	AC1	359358	6	
Studholme, W.	AC2	406050	1	
Sturch, W.J.	AC2	135619	1	
Sturgiss, F.C.	F/Lt		9	Also 36/37 as S/L
Sturman, G.W.	F/O		(3)	30 Squadron
Styles, C.E.	ACl	341035	3, 5	
Sugden, W.G.	LAC	363246	6	
Sullivan, G.G.	AC2	516336	9	
Sullivan, J.	AC2	162313	(3)	
Sutcliffe, H.	AC1	327725	4	
Suter, E.E.	LAC	243678	5	
Sutherland, D.E.	Cpl	349486	6	
Sutherland, W.J.	2/Lt		1, 2, 3	
Sutton, A .	Cpl	731	I	
Sutton, B.E.	G/C		8	
Sutton, F.G.	LAC	509525	9	
Sutton, W.	AC1	351441	4	21/24 in lieu Waz 25
Swain, O.B.	F/Lt		6	
Swaine, S.E.	Cpl	355226	1,6	Medal and Afg 19 to 265957 Pte Royal West Kent Regiment
Swait, F.	Cpl	91202	3	
Swann, B.P.	Sgt	189832	(3)	99 Squadron
Swann, W.	LAC	505467	6	
Swanston, J.R.	F/O		2,3	Deceased
Sweetman, R.L.	LAC	561638	9	Also 36/37
Swift, H.	LAC	510174	8	
Swift, N.	LAC	560408	9	
Swift, R.F.	LAC	561931	9	
Swinton, J.H.	AC2	336986	(5)	N.R.V.I.
Sykes, R.G.R.	LAC	358551	6	P/O (7.2.42)
Symon, G.	Sgt	561932	9	Also 36/37
Tagg, F.T.K.	LAC	353857	6	Also 36/37 as Cpl
Tait, A.C.	ACl	348885	4	
Talbot, J.	F/O		1,3	
Talbot, J.G.	AC1	78560	(1)	
Talkington, J.A,	LAC	366205	9	
Tame, P.H.J.J.T.	LAC	216133	(1)	20 Squadron
Tandon, J.N.	F/O		9	Indian Air Force
Tann, F.H.	Cpl	590038	9	

Name	Rank	No.	Clasp	Comments
Tanner, E.W.	LAC	329528	4	Medal originally allocated for issued to P.A. de Fontenay.
Taplin, C.E.	LAC	364326	6	
Tapp, R.B.	Lt		1,2,3	
Tarrant, L.G.	LAC	507477	6	
Tavendale, J.C.H.	F/Lt		6	Deceased
Taylor, A.	AC2	83199	1	
Taylor, A.E.	AC1	357289	6	
Taylor, C.H.	LAC	362700	6	
Taylor MC, D.E.D.	Lt		1	
Taylor, D.F.	AC2	507532	6	
Taylor, E.	AC1	512913	9	
Taylor, E.R.	LAC	363327	6	
Taylor, E.R.J.	F/Sgt	8128	4	
Taylor, F.V.	AC2	351302	4	21/24 in lieu Waz 25
Taylor, G.J.	Cpl	353963	6	
Taylor, H.	Sgt	224898	1	
Taylor, H.	Cpl	313928	4	
Taylor, H.	LAC	511817	9	
Taylor, H.G.	AM2	89484	1	
Taylor, H.J.	AC1	505172	6	
Taylor, H.V.	AC1	327448	5	
Taylor, J.	AC1	507067	6	
Taylor, J.J.	LAC	512045	9	
Taylor, R.H.	ACl	330644	4	Also took part in Waz 25 Opns
Taylor, V.E.	AC2	261629	1	
Taylor, W.F.H.	AC2	83189	3	
Taylour, G.T.	Cpl	362184	6	
Telfer, J.H.	Cpl	352301	6	
Telling, J.H.	LAC	358489	6	
Temple, J.W.	Cpl	340962	6	
Templeman, F.	LAC	244304	4	21/24 in lieu Waz 25
Terry, G.H.	AC1	194781	(3)	20 Squadron
Terry, V. C.	F/Sgt	86948	8	
Tester, A.G.	LAC	359553	6	
Thacker, A.	AC1	511826	9	
Thackeray, J.W.	LAC	370742	6	
Thatcher, R.F.	LAC	11307	1	31 Squadron
Thatcher, W.	Cpl	346125	6	A/Sgt
Theak, W.E.	F/Lt		4	
Thewlis, L.	AC	350218	(5)	N.R.V.l.
Thom, W.D.	F/Lt		1,(3)	97 Squadron. Medal O.C.
Thomas, A.	AC1	341466	4	
Thomas, E.	Cpl	506719	9	Also 36/37 as Sgt
Thomas, F.M.	Cpl	560932	9	
Thomas, G.W.J.	Sgt	326862	6	
Thomas, H.	Cpl	346002	1	
Thomas, H.C.	AC2	242354	(3)	
Thomas, J.B.P.	F/O		9	Also 36/37
Thomas, J.E.G.	F/Lt		6	
Thomas, K.R.	F/O		(5)	N.R.V.l.
Thomas, L.H.	AC1	515309	9	Also 36/37

Name	Rank	No.	Clasp	Comments
Thomas, R.A.	Cpl	2817	1	
Thomas, R.I.	AC1	515024	9	
Thomas, R.L.	F/Lt		1	31 Squadron
Thomas, W.J.	AC1	332574	2,3	20 Squadron
Thompson, A.E.S.	AC1	340991	4	
Thompson, C.N.	F/O		1	
Thompson, C.R.	AC1	338020	2,3	20 Squadron
Thompson, D.A.	LAC	363020	6	
Thompson, E.S.	Cpl	365061	9	
Thompson, F.J.	LAC	364327	6	
Thompson, H.	Cpl	347788	9	Also 36/37 as Sgt
Thompson, H.	AC2	95837	(1)	5 Squadron
Thompson, J.	LAC	364299	6	
Thompson, J.	AC1	514189	9	
Thompson, L.	LAC	343660	4	
Thompson, S.G.	AM2	404200	1	
Thompson, S.L.	Sgt	346616	6	
Thompson, T.C.	LAC	505542	6	
Thompson, T.F.W.	F/Lt		2,3,4	20 Squadron
Thompson, W.A	AC2	190057	3	
Thomsett, F,	F/Sgt	314406	5	
Thomson, A.	AC1	364307	6	
Thomson, B.C.	LAC	510372	9	
Thomson, D.	F/O		1	31 Squadron
Thomson, D.L.	P/O		5	
Thomson, J.R.	AC1	345055	4	
Thomson, L.	LAC	355273	6	
Thomson, W.P.J.	F/O		6	
Thorn, H.	AC2	343100	3,5	20 Squadron
Thorn, W.D.	F/Lt		1	
Thorne, G.	LAC	359310	6	
Thorne, H.G.	LAC	105019	6	
Thorns, J.F.	LAC	353988	6	
Thorogood,	AM2	405045	1	
Thorp, A.E.G.H.	LAC	364774	6	
Thorpe, A.R.T.	LAC	60778	1	
Thorpe, H.	LAC	362191	6	
Thurston, W.A.	LAC	358864	6	
Thwaite, G.R.	F/O		9	
Thwaites, L.C.	LAC	506511	6	
Tibbles, C.H.	Cpl	362219	8	
Tibbs, W.A.	AC1	362203	6	
Tice, G.H.	LAC	507671	6,9	Also 36/37 as Sgt. Deceased
Tidey, H.G.	Cpl	5817	1	
Tillbrook, E.	AC2	348250	4	21/24 in lieu Waz 25
Tilley, H.	Capt		1,3	52 Wing
Tilley, J.E.	LAC	505376	6,8	
Tills, L.	LAC	370501	6	
Tillson, A.J.	AC2	275772	(1)	5 Squadron
Tilston, J.	AC1	327189	5	
Timmins, J.W.	LAC	356013	6	
Tindal-Carill-Worsley,				

Name	Rank	No.	Clasp	Comments
G.N.E.	F/O		6	
Tingley, H.I.	Sgt	364302	6	
Tinker, W.T.E.	AC2	406332	(1)	20 Squadron
Tipper, C.N.	LAC	326335	4	
Tobin, E.V.	LAC	510534	9	
Todd, F.W.	F/O		5	
Todd, J.W.	LAC	505541	6	
Tolfrey, A.	Cpl	353358	6	
Tolladay, S.	ACl	362190	6	
Tollerfield, J. C.	LAC	561650	9	Also 36/37
Tolley, W.E.	AC1		I	
Tollfree, J.H.	Sgt	16528	1	
Tolliday, A.J.A.	AC1	515535	9	Also 36/37
Tombeur, A,J.	A/Cpl	37286	1	20 Squadron
Tomes, A.E.	LAC	358578	6	
Tomlinson, S.A.	AC1	561651	9	Also 36/37
Tompkins, V.H.	F/Lt		8 , 9	
Toner, E.	LAC	560972	9	Also 36/37
Toner, S G.	AM3	22497	1	
Tonkin, F.J.	LAC	333955	4	99 Squadron
Tonkin, J.	AC1	357343	4	Medal issued by his Regiment
Tonkin, V.	LAC	345405	4 , 9	
Tonkinson, W.	LAC	354695	6	
Toogood, A.E.	LAC	328391	4	
Toogood, C.F.	F/O		1 , 2 , 3	
Toone, C.C.	AMl	78876	1 , 3	
Tooth, F.C.	AC1	288939	(3)	
Toppin, A.	Sgt	107200	4	
Toseland, W.	AC2	135596	1	
Tout, J.H.	ACl	512360	9	
Townend, J.B.	F/O		(5)	N.R.V.I.
Towns, S.T.	F/Sgt	4158	5	
Townsend, A.	LAC	287153	1	20 Squadron
Townsend, A.A.	F/Lt		6	
Townsend, R.E.	LAC	342754	4	
Townsend, W.L.	LAC	309101	6	
Townshend, A.E.	F/Sgt	313936	4	
Townshend, E.F.	Cpl	342683	9	Also 36/37 as A/Sgt
Tracy, C.H.W.	LAC	357765	6	
Travis, G.R.	Capt		1	
Treby, P.J.	Cpl	246607	4	
Tremewan, T.A.	AC1	347852	4	
Trenary, R.	Cpl	365066	6	
Treppass, A.	LAC	336052	(3)	99 Squadron
Trevethick, C.L.	F/Sgt	1062	4	
Trimm, E.G.	Cpl	350669	6	
Tripp, J.E.C.	AC1	358457	(6)	N.R.V.I.
Trott, F.W.	S/L		6	
Truss, J.E.	Lt		1	114 Squadron
Tuck, J.A.H.	P/O		9	
Tucker, E.F.	AC1	350121	4	
Tucker, H.L.	LAC	511519	9	

Name	Rank	No.	Clasp	Comments
Tucker, R.G.	AC1	341712	4	21/24 in lieu Waz 25
Tucker, W.J.	AM2	404266	1	
Tuckey, J.H.	SM2	2389	6	
Tulloch, J.	AC1	342389	4	21/24 in lieu Waz 25
Tumilty, J.	AC1	334368	(2),(3)	N.R.V.I.
Tungay, P.J.	LAC	506165	6	
Turley, A.	Sgt	8905	1	
Turnbull, R.McK.	LAC	359838	6	
Turner, A.T.	LAC	159965	5	
Turner, C.D.	AC2	345375	4	
Turner, C.H.	F/Lt		9	A1so 36/37
Turner, E.W.	AC2	187201	1	
Turner, F.D.	LAC	560707	9	-
Turner, G.C.	Cpl	560532	9	Also 36/37 as Sgt
Turner, H.	AC2	515294	9	Also 36/37 as LAC
Turner, H.H.	AC1	508292	6	
Turner, H.P.	AC2	90866	4	
Turner, J.C.	AM1	404267	1	
Turner, L.S.	AC1	359820	6	
Turner, N.	Cpl	30187	1	
Turner, N.W.	Cpl	560449	6,9	
Turner, T.	LAC	365071	6	
Turner, W.	AC1·	5139S7	9	Also 36/37 as Cpl now S/L
Turney, S.W.	LAC	348776	4	21/24 in lieu Waz 25
Turton, G.S.	AC2	333474	4	
Tuson, R.	LAC	357440	6	
Tuttle, G.W.	F/Lt		9	Also 36/37 as S/L
Tuttlebee, L.A.	Sgt	302889	(1)	31 Squadron
Tweedie, H.A.	F/Lt		1	No.I Indian Wing
Tweedle, J.	LAC	64567	1,3	
Twyman, G.	Sgt	32176	1	
Twyman, S.H.	Cpl	353097	6	
Tye, C.M.	AC1	326336	5	
Tyler, A.L.	LAC	560930	9	
Tyler, E.	AM2	66480	1	
Tyler, E.A.	LAC	362695	6	Also 37/39
Tyrrell, E.W.	O/O		1	
Tyrrell, Z.E.	LAC	366433	9	
Tyson, W.	LAC	512066	9	
Tyssen, J.H.S.	S/L		(3)	20 Squadron
Underhill, R.A.E.	LAC	366220	9	
Underhill, R.E.H.	AC1	335020	4	21/24 in lieu Waz 25
Underwood, J.	AC1	5496	1	
Underwood, J.E.	AM1	445995	1	
Unett, N.	Sgt	55149	2,3	
Upham, A.C.	Capt		1	114 Squadron. Special List
Upton, C.E.	Cpl	337746	(3)	99 Squadron
Upton, C E.	AC1	328298	4	21/24 in lieu Waz 25
Usher, A.	LAC	326854	(5)	N.R.V.I.
Usher-Somers, E.C.	F/0		(3)	20 Squadron

Name	Rank	No.	Clasp	Comments
Utteridge, F.E.	AC1	326872	4	21/24 in lieu Waz 25
Vachell, J.L.	F/Lt		(3)	99 Squadron
Valentine, E.	AC2	336788	3	52 Wing
Valentine, F.S.	LAC	365072	6	Deceased
Valentine, H.E.	Cpl	13564	I	31 Squadron
Van Beest, F.C.L.	LAC	506068	6	
Van Loock, G.E.	LAC	506378	6	
Van Zeller, H.J.	AC1	506438	6,9	
Vardy, K.S.	LAC	350129	6	
Varey, J.	AC1	506682	6	
Varney, R.F.	AC1	289942	(3)	31 Squadron attached 52 Wing
Vasse, G.H.	F/Lt		9	
Vaughan, J.J.	AC2	287854	4	
Veal, H.J.E.	Cpl	358700	6	
Veal, L.W.R.	AC1	508782	9	
Venning, W.H	Sgt	31015	1	
Verney, E.A.J.	AC1	515134	9	Also 36/37 as LAC
Verney, N.W.	LAC	561228	9	Also 36/37
Veryard, R.G.	F/Lt		6,9	
Vetch, W.H.	F/O		5	
Veysey, E.W.	AC1	90238	(3)	
Vickers, F.W.	AC2	247017	(3)	20 Squadron
Vidler, F.E.	LAC	353911	6	
Viggers, G.W.	LAC	359364	6	
Vigor, N.M.	LAC	560499	9	
Viles, E.W.	LAC	365957	9	Also 36/37
Villiers, F.E.E.	F/O		1	31 Squadron
Vincent, C.Mc.C.	F/O		1	31 Squadron
Vincent, F.J.	F/Lt		(3)	20 Squadron
Vincent, R.J.	AC1	512375	9	
Vinson, S.G.	LAC	359687	6	
Vintras, R.E. de T.	F/O		6	
Virgin, H.S.	LAC	513292	9	
Voller, S.B.	LAC	590064	9	
Wade, M.	LAC	326210	4	
Wade, W.G.	AC1	328022	5	
Waddy, E.B.	F/O		8,9	
Wain, E.F.	F/O		6	
Wain, J.S.	LAC	366490	9	Also 36/37 as Cpl
Waite, A.J.	LAC	251488	5	
Wake, W.E.	AC1	370687	6	
Walden, F.N.	LAC	62751	(1)	20 Squadron
Waldock, F.W.	Sgt	407711	6	
Walford, C.S.R.	LAC	244737	4	
Walker, C.	F/O		3,4	28 Squadron
Walker, C.A.	ACI	403625	I	
Walker, E.J.W.	LAC	365031	8,9	
Walker, E.W.	W/O	224836	9	

Name	Rank	No.	Clasp	Comments
Walker, F.R.	LAM	7867	1	
Walker, G.E.	Cpl	507860	9	
Walker, J.	LAC	350820	4	
Walker, J.F.	AC2	342710	(5)	N.R.V I.
Walker, J.W.	Cpl	327132	6	
Walker, S.G.	Sgt	246000	2,3	99 Squadron
Walker, W.D.	AC2	507758	6	
Walker, W.H.	LAC	352206	4	21/24 in lieu Waz 25
Wall, A.	F/O		6	
Wall, T.G.	LAC	362208	6	
Wallace,	Sgt		I	
Wallace, E.S.	AC2	100939	1	
Wallace, J.	AM3	165339	1	Medal issued 10.5.46
Wallace, R.	F/Sgt	337918	9	
Waller, E.R.	Sgt	149293	4	
Waller, G.L.	AC1	356062	6	
Waller, R.R.S.	F/O		4	
Wallington, W.F.	Sgt	364350	6	
Wallis, A.G.	LAC	365335	6	
Wallis, A.T.	Cpl	202472	I	
Wallis, E.J.	LAC	506486	6,9	
Walmsley, A.W.	LAC	327082	4	21/24 in lieu Waz 25
Walmsley, W.A.	Cpl	14038	1	
Walser, A.A.	W/C		5	
Walser MC, J.G.	F/Lt		4	
Walsh, C.	LAC	341541	5	
Walsh, E.	AC1	512863	9	Also 36/37 as Cpl
Walsh, H W.	Sgt	149424	4	21/24 in lieu Waz 25
Walsh, J L.	F/Lt		9	Also 36/37
Walsh, J.T.	AC2	282775	4	
Walsh, L.W.	LAC	314420	5	
Walshaw, A.G.	AC1	314421	4	
Walter, E.J.	LAC	358755	6	
Walter DFC AFC, E.L.A.	F/O		9	Also 36/37 as F/Lt and 37/39 60 Squadron
Walter, G.S.	LAC	351585	6	
Walter, R.	Cpl	361663	6	
Walter, R.S.	F/O		4	
Walters, A.	AM3	407788	1	
Walters, T.H.	LAC	340299	4	
Walters, T.L.	Cpl	330751	(3),(5)	20 Squadron
Walton, W.G.	LAC	363301	6	
Want, R.E.	Sgt	245493	9	
Warburton, J.	F/Lt		6	
Ward, A.C.	AC2	88974	3	
Ward, E.L.S.	F/O		6	Also 37/39 as S/L
Ward, J.	AC1	359625	6	
Ward, J.P.	AC1	246164	2,3	
Ward, K.E.	F/O		2,3	31 Squadron
Ward, L.A.	Sgt	335243	9	
Ward, T.E.	LAC	505983	6	
Warden, W.W.F.	AC1	82498	4	

Name	Rank	No.	Clasp	Comments
Ware, J.C.	LAC	76962	1	
Warehan, A.	ACl	337911	2,3	
Wareing, H.	LAC	362220	6	
Wargent, W.A.	Cpl	355820	6	
Waring, C.L.	LAC	506010	6	
Warman, R.	ACl	506384	6	
Warmer, H.R.	AC1	359677	6	
Warmington, C.G.	AC1	361657	6	
Warnt S.	LAC	349696	5	
Warne, H.	LAC	66582	1,3	20 Squadron
Warner, C.H.T.	F/O		9	Deceased 21.9.36
Warner, E.F.	AM2	406289	1,(3)	
Warner, W.E.	Sgt	359693	9	Also 36/37
Warnett, D.A.	LAC	513667	8	
Warren, T.F.	LAC	370162	6	
Warren, W.A.	LAC	359949	6	
Warren, W.H.F.	LAC	357865	6	
Warrener, G.	F/Sgt	3888	4	21/24 in lieu Waz 25
Warrington, H.F.	LAC	343991	4	
Warth, H.	Sgt	340545	6	Deceased. Medal used for 335700 R.R.G. Morris
Waters, N.E.	AC1	406620	1	
Watkins, A.L.	F/O		1	
Watkins, R.C.	LAC	362715	6	
Watkins, R.H.J.	Cpl	34S547	9	
Watkins, V.E.M.	Cpl	366464	9	Also 36/37 as Sgt
Watkins, W.R.	AM2	200994	1	
Watkinson, C.	LAC	342818	5	
Watkinson, E.J.	Sgt	19301	1	
Watson, A.C.	F/O		6	
Watson, J.	AC2	332454	2,3	20 Squadron
Watson, J.S.F.	O/O		1	
Watson, R.	2/Lt		1,6	30/31 as 359702 LAC
Watson, R.G.	F/O		9	
Watson, W.	LAC	505360	9	
Watt, R.C.	AM1	133521	1	
Watton, J.	LAC	15284	1,3	
Watts, A.I.	2/Lt		1	
Watts, E.	AC1	348973	4	
Watts, E.W.	LAC	359987	6	
Watts, F.J.	F/Lt		1,2,3,4	31 Squadron att. 20 Sguadron
Watts, H.	LAC	35510	1	Aden Flight
Watts, J.J.	F/O		9	27 Squadron
Watts, R.A.	LAC	326829	4	
Watts, R.E.	F/O		6	
Watts, V.B.	AC1	329503	6	N.R.V.I.
Watts-Read, A.M.	F/O		6,8	
Way, D.T.	F/Sgt	362760	9	
Way, H.	LAC	590243	9	Also 36/37 as Cpl
Waybourne, T.R.	Cpl	17090	1	
Wayland, C.F.	Sgt	348002	9	Also 36/37
Waymouth, O.S.	Lt		2,3	

Name	Rank	No.	Clasp	Comments
Weait, C.J.L.	LAC	347154	4	
Weaver,	AM2	22479	1	
Webb, A.E.	AC1	145516	5	
Webb, C.F.	LAC	341200	5	
Webb, E.J.	Cpl	335756	9	Also 36/37 as Sgt
Webb, F.	LAC	510033	9	Also 36/37 as Sgt
Webb, F.G.	LAM	78886	1	
Webb, H.A.	AC1	345345	4	
Webb, R.J.A.	Cpl	5951	1,3	
Webb, W.	F/Sgt	314431	4	
Webb-Bowen, T.I.	A/Cdr	A/C	2,3	Commands R.A.F.
Webber, A.H.	LAC	365979	6	
Webber, C.H.	AC2	160637	(1)	Special List
Webber, G.R.	LAC	364372	6	
Webster, C.E.	Cpl	505415	9	
Webster, S.N.	F/O		3,4	28 Squadron & 60 Squadron
Webster, T.	LAC	370697	6	
Webster, W.	SM2	2436	4	
Wedgbury, F.	LAC	561408	9	
Weeden, A.J.	LAC	363230	6	
Weeden, J.E.	AC1	506979	6	
Weedon, C.W.	F/O		5	
Weedon, R.A.M.	AC1	514020	9	Also 36/37 and 37/39
Weeks, H.R.	LAC	561667	9	
Weeks, W.C.	AM1		1	
Weeks, W.F.	AC1	266426	4	
Weeks, W.J.	AC2	335552	4	
Weir, H.	Sgt	280069	4	
Weir, W.F.	AC2	332008	1,3,4	Medal O.C.
Welbourne, W.F.	Cpl	340793	5	
Welch, C.D.G.	F/O		6	
Welch, L.	Sgt	343269	8	
Welch, S.R.	LAC	359446	6	
Welch, S.T.	LAC	362209	6	N.R.V.I.
Wellbeloved, S.W.	LAC	254121	4	
Wells, C.R.	AC1	347236	4	
Wells, F.J.	AC2	245042	1	
Wells, F.T.	LAC	358194	6	
Wells, P.G.	F/O		1	31 Squadron
Wells, R.J.	AC1	342400	4	
Wells, T.	AC	405776	1	
Wells, W.C.	LAC	362212	(6)	N.R.V.I.
Welsh, A.H.	AC1	335174	4	21/24 in lieu Waz 25
Welsh, W.	AC2	329445	3	20 Squadron
Welton, F.C.	LAC	159078	4	
Wesson, R.E.	LAC	366235	9	Also 36/37 as Cpl
West, H.	AC2	512771	8,9	Also 36/37 as LAC
West, J.	Cpl	346051	9	Also 36/37 as Sgt
West, J.	AC1	330634	5	
West, P.R.	AC2	511590	6,9	
West, R.F.	Cpl	143742	6	
Westaway, H.W.	F/O		4	

Name	Rank	No.	Clasp	Comments
Westbrook, E.C.	LAC	155700	4	
Westbrook, F.G.	LAC	364357	9	
Westcott, C.W.	Cpl	89172	6	
Western, J.G.	F/O		5	
Weston, H.C.	LAC	363041	6	
Weston, J.G.W.	F/O		6	
Westwood, J.	AC1	335205	5,6	
Whalebone, G.H.	LAC	562648	9	Also 36/37
Whaley, G.A.J.	LAC	564431	9	Also 36/37
Wharton, A.	SM1	4203	1	
Wharton, A.L.	AC2	409042	1	
Whatling, G.	Cpl	590137	9	Also 36/37 as Sgt. Now officer
Wheatley, D.H.	LAC	590132	9	
Wheatley, G.V.	F/O		5	Deceased
Wheatley, S.J.	AC1	507553	6	
Wheaton, A.R.	LAC	363047	6	
Wheeler, A.	AM1	407598	1	
Wheeler, H.A.	AC2	328411	(5)	N.R.V.I.
Wheeler, H.G.	F/O		6,9	Also 36/37 as S/L
Wheeler, H.J.	F/Sgt	11256	3	
Wheeler, J.W.T.	LAC	505497	6	
Wheldon, J.F.	LAC	157380	(3)	99 Squadron
Whitaker, A.	LAC	334123	4	
Whitbread, E.A.	LAC	365818	6	
White,	AC1	62145	1	
White, A E.	SM2	80436	6	
White, A.E.	LAC	328384	4	21/24 in lieu Waz 25
White, A.W.	LAC	505953	6,8	
White, C.G.N.	Cpl	560800	9	
White, F.	F/Sgt	351871	9	
White, H.	LAC	359273	6	
White, J.W.	LAC	363044	6	
White, L.G.	Sgt	110761	1	48 Squadron
White, P.G.	P/O		1	31 Squadron
White, R.A.L.	LAC	561664	9	
White, R.H.	AC2	120644	1,2,3	
White, R.W.	F/Lt		6	
White, V.G.B.	Cpl	366230	8,9	Also 36/37 as Sgt
White, W.A.	AC1	511046	9	
Whitehead, B.C.	AC2	105458	1	
Whitehead, F.N.	F/O		(1),2,3	
Whitehead, R.P.	S/L		4	
Whitewood, L.A.	LAC	514264	9	
Whiting, F.E.	LAC	338169	6	
Whiting, H.F.	AC1	332300	4	
Whittaker, J.T.	S/L		4	
Whitton, R.	F/Sgt	4330	6	
Whitwell, J.M.	Sgt	363271	6	
Why, H.W.	AC2	510466	6,9	
Whyte, A.	LAC	334235	3	
Wickerson, C.W.	Cpl	354266	6	
Wickham, E.G.W.	AC2	155656	(1)	20 Squadron

Name	Rank	No.	Clasp	Comments
Wickham, S.F.W.	LAC	364353	9	
Wicks, F.F.	P/O		6	
Widmer, R.C.	LAC	330605	(5)	N.R.V.I.
Widocks, R.	Cpl	248753	4	
Wiggins, W.	Sgt	7828	1	
Wilce, F.E.	Cpl	347581	6	
Wild, A.J.	Sgt	363743	6	
Wildbore, R.M.	W/O	963	9	
Wilderspin, J.C.	Sgt	245220	3	
Wileman, W.G.	Cpl	560788	9	Also 36/37
Wilkinson, D.E.	LAC	512102	9	Also 36/37 as Cpl
Wilkinson, V.J.	LAC	333873	3	99 Squadron
Wilks, C.W.	AC2	276422	3	20 Squadron
Wilks, F.H.	LAC	505493	6	
Willcox, A.L.	O/O		1	
Willcox-Jones, D.S.	AC1	363106	6	
Willers, W.H.	Cpl	251489	4	
Willes, J.	LAC	505755	6	
Williams, A.	Cpl	326058	4	
Williams, A.B.E.	F/Sgt	351867	9	
Williams, A.E.	AC2	306589	(3)	
Williams, C.C.D.	F/O		6	
Williams, C.R.	LAC	363763	6	
Williams, D.R.	LAC	364339	6	
Williams, E.	LAC	364393	6 , 8	
Williams, E.	AC1	345765	5	
Williams, E.E.R.	Sgt	82908	6,8,9	Also 36/37 as F/Sgt and 37/39
Williams, F.T.	LAC	365037	9	
Williams, G.	Cpl	342801	4	
Williams, G.A.	S/L		(3)	20 Squadron
Williams, G.H.	LAC	356942	(6)	N.R.V.I.
Williams, G.T.	AC1	562950	9	
Williams, J.F.	Cpl	115767	4	
Williams, R.C.	2/Lt		1	31 Squadron
Williams, R.G.	LAC	506503	6	
Williums, S.	LAC	561671	9	Also 36/37
Williams, S.T.	AC1	511104	9	
Williams, W.P.	AC1	356795	6	
Williamson, L.	LAC	159949	5	
Williamson, P.C.	Cpl	336024	4	
Williamson, W.	LAC	187570	(3)	52 Wing
Willis, F.C.	AM2	410170	1	
Willis, G.C.S.	LAC	364347	6	
Willmott, B.H.	LAC	365562	8	
Willmott, H.V.	Sgt	356159	9	
Willott, C.C.	F/Sgt	362231	9	Also 36/37.Commissioned
Willoughby, N.	LAC	362718	6	
Wills, E.R.	Cpl	509154	9	Also 36/37 as Sgt
Wilsdon, A.A.	AC2	510613	6	
Wilson, A.B.	AC	407600	1	114 Squadron
Wilson, A.T.	LAC	509026	8	
Wilson, C.W.	Cpl	343011	5	

Name	Rank	No.	Clasp	Comments
Wilson, D.A.R.	Cpl	361651	6	
Wilson, E.	LAC	346950	4	
Wilson, E.J.G.	LAC	359237	6	
Wilson, F.	LAC	511190	8,9	
Wilson, F.A.	AC1	510984	9	
Wilson, F.H.	AC2	97031	1	
Wilson, J.A.	Lt		1	
Wilson, M.W.	AC1	247707	(3)	20 Squadron
Wilson, R.G.	LAC	590532	9	Also 36/37 asCpl
Wilson, R.W.	Cpl	348372	6	
Wilson, W.F.	S/L		4	
Wilton, G.N.	F/O		( )	
Wiltshire, C.S.	Cpl	364367	6	Deceased
Wiltshire, L.A.	Cpl	326180	(5)	N.R.V.I.
Wiltshire, W.F.	AC1	350603	4	
Wiltshire, W.P.	F/O		5	
Winch, H.E.	P/O		2,3	
Windsor, W.H.	LAC	353244	6	
Winkler, H.R.W.	AC1	564597	9	
Winstanley, G.	F/O		1	
Winter, F.E.	SM2	343	(3)	97 Squadron
Winter, W.	AC2	149664	1	31 Squadron
Wintour, J.W.	Cpl	25190	1	
Wise, C.H.	LAC	505501	6	
Wise, G.E.	AC1	348199	5	
Wiseman, T.L.	Sgt	78925	1	
Wiskin, E.W.	AM2	132837	1	-
Witcher, W.	Sgt	8437	1	
Witheridge, S.P.	LAC	354863	6	
Withers, S.A.	AM1	404298	1	
Wodhams, W.J.	AC2	408849	1,2,3	
Wolsworth, S.	AM2	13770	1	
Wood, A.C.	LAC	361747	6,9	
Wood, A.J.	AM1	47137	1	
Wood, C.H.I.	LAC	370236	6	
Wood, D.H.G.	F/O		6	Deceased
Wood, E.	AC1	66311	(1)	20 Squadron
Wood, F.C.	Cpl	311023	4	
Wood, F.E.	Cpl	365813	6	
Wood, J.	Sgt	82587	1,3	
Wood, J.	AM2	140897	1	
Wood, L.G.A.	AC2	314444	4	
Wood, S.E.	AC2	327021	5	
Wood, W.	AC2	334039	4	
Wood, W.E.J.	AC2	141221	1	31 Squadron
Wood, W.G.S.	F/O		8	
Wood, W.S.	LAC	347527	5	
Woodcock, W.F.	Sgt	350682	6	
Woodhouse, C.J.	AC1	508783	6,9	
Woodhouse, G.P.	F/O		9	
Woodhouse, W.J.	AC2	277318	1,2,3	
Woodley, S.C.	LAC	341607	5	

Name	Rank	No.	Clasp	Comments
Woods, A.G.	Cpl	291808	5	Also 36/37 as F/Sgt
Woods, E.V.	LAC	370942	6	
Woods, J.	AC1	346147	(5)	
Woods, J.	Cpl	65087	4	36/37 as W/O. Now F/O
Woods, P.	Cpl	348995	9	
Woods, S.F.	LAC	365794	9	Also 36/37 as Cpl
Woods, W.	AC1	1635063	4	Formerly 334773
Woodthorpe, R.C.	F/Sgt	81374	8	
Woodville, M.E.	AC1	334326	3	
Woodward, A.	LAC	512696	9	
Woolcott, L.F.	Cpl	362729	6	
Woolgar, R.A.	LAC	362210	6	
Woollard, J.R.	F/Sgt	313303	5	
Woolnough, D.R.	AC1	506286	6,8,9	
Wooltorton, G.E.H.	F/Sgt	314934	6	
Worbey, H.C.	AC2	292225	(3)	
Worth, L.G.	LAC	362710	6	
Worton, A.	LAC	366470	9	
Wragg, E.	AC2	343438	( )	Medal named ACII Wragge
Wrangham, J.	AC2	148729	1	
Wray, A.M.	F/O		(3),(4)	20 Squadron
Wren, E.F.	LAC	347273	4	
Wren, J.E.	Sgt	345850	6	Deceased
Wrench, F.W.	F/O		4	
Wright, C.L.Y.	P/O		9	Also 36/37 and 37/39. Deceased
Wright, F.J.	AC2	134876	1	
Wright, H.	F/Sgt	335688	9	
Wright, H.	LAC	363311	8	
Wright, H.C.	Sgt	36464	(1)	48 Squadron
Wright, H.F.	LAC	247539	4	21/24 in lieu Waz 25
Wright, H.M.S	F/Lt		9	Also 36/37 as S/L
Wright, R.	LAC	363282	6	Not entitled to Mohmand 1933
Wright, S.G.	LAC	327585	4	21/24 in lieu Waz 25
Wright, W.	P/O		1	
Wright, W.	AC1		1	
Wright, W.J.H.	AC1	515231	9	Also 36/37
Wright, W.W.	Cpl	347181	6,8	
Wyatt, A.E	LAC	560470	9	
Wyatt, G.	LAC	358519	6	
Wyatt, W.G.	AC1	149760	1	20 Squadron
Wycherley, L.	Lt		1	
Wynn, O.	LAC	149162	6	
Wynter-Morgan, W.	P/O		6	
Yarnold, F.A.	AC2	197023	(1)	27 Squadron
Yarwood, A.E.	LAC	358176	6	
Yeoman, J.	Cpl	157962	7	
Yeomans, A.	LAC	326059	4	
Yorke, T.E.	LAC	510582	9	Also 36/37 as Sgt
Youings, A.W.	Cpl	560220	9	
Young, E.	AC1	508883	8,9	

Name	Rank	No.	Clasp	Comments
Young, F.	AC1	370753	6	
Young, H.C.	Lt		1	
Young, J.C.	AM2	406118	1	
Young, J.F.	AC1	240771	1,(3)	99 Squadron
Young, R.	Sgt	333583	6	
Young, R.A.	LAC	358218	6	
Young-Evans, E.T.	AC2	359495	6	
Younger, D.W.	Sgt	405480	1	
Yule, T.	Cpl	344926	9	Also 36/37